FOLLOW IN THE FUTURE WITH THE SHADOW

INTERNATIONAL INTERGALACTICAL SPACE FEDERATION (IISF)

GEORGIY SERGEYEVICH GARBUZ

Published 2024
Printed in the United States of America

First Edition
ISBN (softcover): 978-1-963380-18-7
ISBN (e-book): 978-1-963380-19-4

For information, address:

Holzer Books LLC
8 The Green, Ste. A
Dover, Delaware 19901 USA

For information about special discounts available for bulk purchases, sales promotions, and educational needs, contact:
info@holzerbooksllc.com
+1 (888) 901-7776

TABLE OF CONTENTS

INTRODUCTION

Outer space is perhaps the most secret and unexplored place in the world. All countries and scientists endeavor to explore what lies beyond. Since the historic moon landing by the Soviet Union, it has become increasingly clear what humanity can achieve and potentially inhabit in other planets. People can realize and resolve many of their problems, and even expand Earth's territory.

Throughout the years, it has been a race among countries, each vying to be the first. Scientists have tirelessly worked, creating technologies and envisioning new possibilities for when space exploration becomes a reality.

The idea of all countries coming together to form an International Intergalactic Space Federation is indeed ambitious and holds great potential. This collaborative effort would enable nations to pool resources, share technologies, and collectively explore the vast unknown of outer space.

By establishing an international government led by influential countries like America, Russia, Germany, India, China, and England, the vision of exploration and expansion beyond Earth's boundaries can be realized. Each nation would maintain its independence while contributing as members and partners of the Federation.

The Space Government's primary purpose would be to foster cooperation among nations, facilitate space exploration, and discover new territories and resources for the benefit of all humanity. This initiative could serve as a unifying force, redirecting focus from conflicts and wars on Earth to the pursuit of knowledge and progress in space.

The current global situation, marked by conflicts and wars driven by power struggles and resource disputes, underscores the urgent need for collaboration and unity. Events like the Russian-Ukrainian war highlight the devastating consequences of geopolitical tensions, affecting not only the involved nations but also causing ripple effects across the world.

Through the International Intergalactic Space Federation, countries could channel their energies and resources towards peaceful exploration and advancement, alleviating suffering and fostering prosperity for all. By transcending earthly conflicts and focusing on the boundless opportunities of space, humanity can take a significant step towards a brighter and more harmonious future.

The current global economic challenges, compounded by restrictions and sanctions, emphasize the importance of international partnerships and cooperation. In the face of inflation and rising prices, nations are realizing the necessity of working together to navigate through crises.

Events like the Russian-Ukrainian war serve as stark reminders of the unpredictability of the future and the need for preparedness. It underscores the importance of ensuring that countries have sufficient supplies to meet the needs of their populations.

The project of establishing the International Intergalactic Space Federation encompasses numerous initiatives aimed at fostering peace and prosperity worldwide. One such initiative is Project Lyubov, which aims to promote love and reconciliation between Russia and Ukraine, thereby contributing to global harmony.

The International Intergalactic Space Federation is envisioned as an international government, with each country of the world being allies. The founding nations, including America, Russia, Germany, India, China, and England, possess advanced space research capabilities and have already made significant contributions to space exploration and technology development.

These nations have invested resources in space research, leading to innovations such as smartphones, internet technology, and more. By leveraging their expertise and resources, they are poised to lead the way in establishing the new government alliance.

Ultimately, the goal of the International Intergalactic Space Federation is to promote peace, prosperity, and exploration beyond Earth's boundaries. Through collaboration and shared efforts, humanity can aspire to a future where the benefits of space exploration are enjoyed by all.

The Space Federation is divided into six territorial sections, each controlled by one of the founding countries, encompassing a total of thirty-two countries within these territories. This business project plan outlines how the territories are assigned to the founders and the governance structure for controlling and exploring these regions of space.

INTERNATIONAL INTERGALACTICAL SPACE FEDERATION (IISF)

The International Intergalactic Space Federation government is designed to secure and research space. It was founded by the USA, Russia, India, England, Germany, and China, and includes participation from 195 countries around the world, united in their shared vision for Earth's future. As a space government, the IISF is responsible for safeguarding space territory, conducting research, expanding territory through planetary colonization, and seeking existing life across the universe. Their mission also includes improving living conditions on various planets, providing job opportunities and businesses to the planetary populations, among other objectives.

The International Intergalactic Space Federation will collect 6% in federation taxes and 1.89% in sales taxes to support an active space research military program spanning the globe and the galaxy.

The head of the IISF, a President elected for a 10-year term, also serves as the Commander of the Chief International Intergalactic Space Agency. This governance structure includes representation from 195 countries with one Senator from each country, serving 8-year terms. The Space Agency boasts a workforce of 47,000,000 people, supported by 75,000,000,000 agents responsible for providing comprehensive information to the Space Government and ensuring the safety and protection of its inhabitants.

The IISF is dedicated to protecting space borders and conducting research. It refrains from participating in local conflicts between countries within planets unless they pose a threat to world or planetary destruction. In such cases, the IISF provides humanitarian aid, offering safety, food transportation, and other forms of support.

The IISF values religious freedom, with Catholic Christianity serving as the main church religion within the federation, offering much-needed spiritual guidance to humanity.

The founder of this visionary project is Georgiy Sergeyevich Garbuz, and the project's revenues will be allocated to healthcare, military agency expenses, space project research, global education initiatives, support for underserved youth, and the protection of space borders. Funding for educational grants will be sourced from taxes on Georgiy Sergeyevich Garbuz's scientific work and other related projects.

The Project For The Future (G.S. Garbuz) was initiated by Georgiy Sergeyevich Garbuz in 2007 and is being carried forward by the National Garbuz Space Academy Trust Corporation.

NATIONAL GARBUZ SPACE ACADEMY TRUST CORPORATION

Ownership:

- 40% Georgiy S. Garbuz, Founder
- 5% Kazakhstan (Government)
- 5% Russian (Government)
- 5% Chinese (Government)
- 5% Germany (European Union Government)
- 5% Indian (Government)
- 5% England (Government)
- 5% United States (Government)
- 5% NASA
- 20% Imperial Space Federation Government (USA)

NATIONAL GARBUZ SPACE ACADEMY TRUST CORPORATION OF IMPERIAL RUSSIA

Ownership:

- 50% National Garbuz Space Academy Trust Corporation (Georgiy Sergeyevich Garbuz)
- 50% National Garbuz Space Academy Trust Corporation of Imperial Russia

Territory Responsibilities:

- Imperial Russia has been allocated 32 countries for the space projects. Security for these projects will be provided by the respective authorities of each country involved.

PROJECT MAMA

Project MAMA is a comprehensive initiative aimed at the construction and expansion of cities in Russia and educational institutions associated with the Garbuz Space Academy. This project focuses on the care, education, training, and future prospects of orphaned children (AA, A1, A2) and children from military backgrounds (A3) in Russia and the CIS countries.

These children will have opportunities for living, studying, and participating in a range of projects, including "Children of the Future," "Mama," "Brother," "Vatican," and "Illiaja." Furthermore, the project aims to ensure the safety and protection of all participants.

In later phases, some participants may have the opportunity for relocation to well-to-do European countries such as England, Germany, Spain, Greece, with exceptions for certain prohibited countries in Africa and the Middle East, including Saudi Arabia.

Another phase involves potential relocation to Antarctica for work, education, training, and participation in projects like "Antar 5," "Children of the Future," "Mom," "Brother," "Vatican," and others.

Additionally, there is a provision for potential relocation to European countries for continued work, education, and training of project participants. Special considerations and monitoring are in place for the protection of these children, particularly in the event of unforeseen circumstances.

NATIONAL GARBUZ SPACE ACADEMY TRUST CORPORATION OF KAZAKHSTAN

Ownership:

- 50% National Garbuz Space Academy Trust Corporation (Georgy Sergeyevich Garbuz)
- 40% National Garbuz Space Academy Trust Corporation of Kazakhstan
- 10% National Garbuz Space Academy Trust Corporation of Imperial Russia

PROJECT PAPA

Expansion and development of the city of Esik, along with the establishment of Garbuz Space School Academies and Garbuz Space Academies, involving the relocation and integration of orphaned children (AA) (A1) (A2) and children of military personnel in Kazakhstan (A3) (inclusive – to be personally verified).

The purpose is to provide these individuals with living arrangements, education, upbringing, training, work opportunities, and involvement in projects such as "PAPA," "Children of the Future," and "MAMA."

The subsequent plan involves their transfer to affluent countries, like England, Germany, Spain, and other European countries. The relocation is intended for further education, training, work, and participation in projects like "Children of the Future," "Brother," "Vatican," and "Illiaja," ensuring the protection of the participants.

Additionally, a phase involves their transfer to Antarctica for work, education, training, and participation in projects such as "Antar 5," "Children of the Future," "Mama," "Brother," "Vatican," and "PAPA."

Special supervision is enforced for the children, and in the event of any harm, the legal provision "For Children" under Article 58, including charges of terrorism, confiscation of both registered and unregistered property, and execution with a special cartridge, is applicable.

NATIONAL GARBUZ SPACE ACADEMY TRUST CORPORATION OF CHECHNYA

Ownership:

- 50% National Garbuz Space Academy Trust Corporation (Georgy Sergeyevich Garbuz)
- 40% National Garbuz Space Academy Trust Corporation of Chechnya
- 10% National Garbuz Space Academy Trust Corporation of Imperial Russia

PROJECT CHECHENECH

Enrollment of orphaned children (A1) from Chechnya, Russia, both before and after their integration into Russia, is undertaken for the purposes of education, upbringing, training, work, and participation in Project MAMA. Subsequently, there is a planned transfer to specific countries, excluding those forbidden (Africa, East, except Saudi Arabia and Iran), with the aim of participating in Project MAMA. This initiative also ensures the protection of participants in projects such as "Children of the Future" and "MAMA," with strict measures in place, including the use of a special cartridge in case of any threats.

NATIONAL GARBUZ SPACE ACADEMY TRUST CORPORATION OF BELARUS

Ownership:

- 50% National Garbuz Space Academy Trust Corporation (Georgy Sergeyevich Garbuz)
- 40% National Garbuz Space Academy Trust Corporation of Belarus
- 10% National Garbuz Space Academy Trust Corporation of Imperial Russia

PROJECT POTATO

The development and expansion of a city in Belarus (Russia) involve the establishment of Garbuz Space School Academies and Garbuz Space Academies, with a focus on utilizing orphaned children (A1) from Belarus for their education, upbringing, training, work, and participation in various projects, including "Mama," "Potato," "Chernovtsi," and others associated with the overarching initiative "Children of the Future."

Following this phase, there is a planned transfer of these individuals to affluent European countries (specific countries to be determined), excluding prohibited regions such as Africa and the East. The primary objective is their continued participation in the "Mama" project, along with ensuring the safety and protection of participants engaged in both the "Mama" and "Children of the Future" projects.

NATIONAL GARBUZ SPACE ACADEMY TRUST CORPORATION OF UZBEKISTAN

Ownership:

- 50% National Garbuz Space Academy Trust Corporation (Georgy Sergeyevich Garbuz)
- 40% National Garbuz Space Academy Trust Corporation of Uzbekistan
- 10% National Garbuz Space Academy Trust Corporation of Imperial Russia

PROJECT AMU DARYA

The undertaking involves the construction and expansion of a city in Uzbekistan (Russia), employing orphaned children (A1) from Uzbekistan, both before and after joining Russia. The objective is to provide them with education, upbringing, training, work opportunities, and engagement in various projects such as "Amu Darya," "Papa," "Mama," "Potato," "Chernivtsi," and other initiatives associated with the overarching theme of "Children of the Future."

Following this phase, there is a planned transfer of these individuals to well-established countries like Turkey and Egypt, with specific countries to be determined. Notably, prohibited regions such as Africa are excluded. The purpose of the transfer is to facilitate their participation in projects like "Pharaoh," "Children of the Future," "Mom," the "Mulla" Project, while ensuring the safety and protection of participants engaged in projects like "Mama," "Children of the Future," and "Papa."

The children are placed under special supervision, anticipating any unforeseen events. Legal provisions, including Article 58 with charges of terrorism, confiscation of both registered and unregistered property, and execution with a special cartridge, are applicable in case of harm to the children.

NATIONAL GARBUZ SPACE ACADEMY TRUST CORPORATION OF KYRGYZSTAN

Ownership:

- 50% National Garbuz Space Academy Trust Corporation (Georgy Sergeyevich Garbuz)
- 40% National Garbuz Space Academy Trust Corporation of Kyrgyzstan
- 10% National Garbuz Space Academy Trust Corporation of Imperial Russia

PROJECT ISSYK-KUL

The initiative involves the construction and expansion of a city in Kyrgyzstan (Russia), establishing Garbuz Space School Academies and Garbuz Space Academies, utilizing orphaned children (A1) from Kyrgyzstan, both before and after joining Russia. The primary aim is to provide these children with education, upbringing, training, work opportunities, and involvement in projects such as "Mom," "Issyk-Kul," "Amu Darya," "Illayazha," "Potato," "Chernivtsi," and other initiatives associated with the overarching theme of "Children of the Future."

Subsequently, there is a planned transfer of these individuals to well-established countries like India, Turkey, and Egypt, with specific countries to be determined. Notably, excluded regions include prohibited areas such as Africa. The purpose of the transfer is to facilitate their participation in projects like "Pharaoh," "Children of the Future," "Mom," the "Mulla" Project, while ensuring the safety and protection of participants engaged in projects like "Mom," "Children of the Future," and "Papa."

The children are placed under special supervision to address any potential risks. Legal provisions, including Article 58 with charges of terrorism, confiscation of both registered and unregistered property, and execution with a special cartridge, are applicable in the event of harm to the children.

NATIONAL GARBUZ SPACE ACADEMY TRUST CORPORATION OF AZERBAIJAN

Ownership:

- 50% National Garbuz Space Academy Trust Corporation (Georgy Sergeyevich Garbuz)
- 40% National Garbuz Space Academy Trust Corporation of Azerbaijan
- 10% National Garbuz Space Academy Trust Corporation of Imperial Russia

PROJECT GYZ GALASY

The endeavor encompasses the construction and expansion of a city in Azerbaijan (Russia), involving the utilization of orphaned children (A1) from Azerbaijan, both before and after their integration into Russia. The primary objective is to provide these children with education, upbringing, training, work opportunities, and engagement in projects such as "Gyz Galasy," "Mama," "Issyk-Kul," "Amu Darya," "Papa," "Potato," "Chernovtsi," and other initiatives related to the overarching theme of "Children of the Future."

Following this phase, there is a planned transfer of these individuals to countries like India, Iran, Egypt, and Indonesia, with specific countries to be determined. Notably, excluded regions include forbidden areas such as Africa. The purpose of the transfer is to facilitate their participation in projects like "Pharaoh," "Children of the Future," "Mom," "Mulla," "Gyz Galasy," and to ensure the protection of participants engaged in projects like "Mom," "Children of the Future," "Illayazha."

Subsequently, there is another planned transfer to the countries of Brazil, where they will participate in the project and work, ensuring the continued protection of participants engaged in projects like "Mom," "Children of the Future," "Papa," and "Hunny." The children remain under special supervision in case of unforeseen events, with legal provisions such as Article 58 in place, including charges of terrorism, confiscation of both registered and unregistered property, and execution with a special cartridge, in the event of harm to the children.

NATIONAL GARBUZ SPACE ACADEMY TRUST CORPORATION OF MONGOLIA

Ownership:

- 50% National Garbuz Space Academy Trust Corporation (Georgy Sergeyevich Garbuz)
- 40% National Garbuz Space Academy Trust Corporation of Mongolia
- 10% National Garbuz Space Academy Trust Corporation of Imperial Russia

PROJECT GOBI

The development and expansion of a city in Mongolia (Russia) involve the utilization of orphaned children (A1) from Mongolia, both before and after their integration into Russia. The primary purpose is to provide these children with education, upbringing, training, work opportunities, and involvement in projects such as the Gobi project, "Giz Galasy," "Mama," "Issyk-Kul," "Amu Darya," "Papa," "Potato," "Chernovtsi," and other initiatives related to the overarching theme of the "Children of the Future" project.

Following this phase, there is a planned transfer of these individuals to countries like India, Iran, Egypt, and Indonesia, with specific countries to be determined. Notably, excluded regions include forbidden areas such as Africa. The purpose of the transfer is to facilitate their participation in projects like "Pharaoh," "Mom," the "Mulla," "Gyz Galasy" project, and to ensure the protection of participants engaged in projects like "Mom," "Children of the Future," "Papa."

Subsequently, there is another planned transfer to the countries of India and China, where they will participate in the project and work, ensuring the continued protection of participants engaged in projects like "Mama," "Children of the Future," "Papa," "Hunny," and "Jochi." The children remain under special supervision in case of unforeseen events, with legal provisions such as Article 58 in place, including charges of terrorism, confiscation of both registered and unregistered property, and execution with a special cartridge, in the event of harm to the children.

NATIONAL GARBUZ SPACE ACADEMY TRUST CORPORATION OF TURKMENISTAN

Ownership:

- 50% National Garbuz Space Academy Trust Corporation (Georgy Sergeyevich Garbuz)
- 40% National Garbuz Space Academy Trust Corporation of Turkmenistan
- 10% National Garbuz Space Academy Trust Corporation of Imperial Russia

PROJECT DEHISTAN

The undertaking involves the construction and expansion of a city in Turkmenistan (Russia), incorporating the use of orphaned children (A1) from Turkmenistan, both before and after their integration into Russia. The primary focus is to provide these children with training, education, upbringing, work opportunities, and participation in projects such as "Dehistan," "Gobi," "Gyz Galasy," "Mama," "Issyk-Kul," "Amu Darya," "Papa," "Potato," "Chernovtsi," and other initiatives linked to the overarching theme of the "Children of the Future" project.

Following this phase, there is a planned transfer of these individuals to countries such as Turkey, Iran, Egypt, and Indonesia. The purpose of the transfer is to facilitate their involvement in projects like "Pharaoh," "Mom," the "Mulla," and the "Gyz Galasy" project, ensuring the protection of participants engaged in projects like "Mom," "Children of the Future," and "Illayazha."

Subsequently, there is another planned transfer to the countries of India and China, where they will participate in the project and work, while ensuring the continued protection of participants engaged in projects like "Mama," "Children of the Future," "Papa," "Hunny," and "Jochi." The children remain under special supervision in case of unforeseen events, with legal provisions such as Article 58 in place, including charges of terrorism, confiscation of both registered and unregistered property, and execution with a special cartridge, in the event of harm to the children.

NATIONAL GARBUZ SPACE ACADEMY TRUST CORPORATION OF ARMENIA

Ownership:

- 50% National Garbuz Space Academy Trust Corporation (Georgy Sergeyevich Garbuz)
- 40% National Garbuz Space Academy Trust Corporation of Armenia
- 10% National Garbuz Space Academy Trust Corporation of Imperial Russia

PROJECT CASCADE

The undertaking involves the construction and expansion of a city in Armenia (Russia), establishing Garbuz Space School Academies and Garbuz Space Academies, with the incorporation of orphaned children (A1) from Armenia, both before and after their integration into Russia. The primary goal is to provide these children with education, upbringing, training, work opportunities, and participation in projects such as "Mama," "Papa," and other initiatives associated with the overarching theme of "Children of the Future."

Following this phase, there is a planned transfer of these individuals to countries like Portugal and other European nations. Notably, excluded regions include forbidden areas such as Africa. The purpose of the transfer is to facilitate their participation in projects like "Mom," "Pharaoh," ensuring the protection of participants engaged in projects like "Mom," "Children of the Future," and "Illayazh."

Subsequently, there is another planned transfer to the countries of India and China, where they will participate in the project and work, while ensuring the continued protection of participants engaged in projects like "Mama," "Children of the Future," "Papa," "Hunny," and "Jochi." The children remain under special supervision in case of unforeseen events, with legal provisions such as Article 58 in place, including charges of terrorism, confiscation of both registered and unregistered property, and execution with a special cartridge, in the event of harm to the children.

NATIONAL GARBUZ SPACE ACADEMY TRUST CORPORATION OF GEORGIA

Ownership:

- 50% National Garbuz Space Academy Trust Corporation (Georgy Sergeyevich Garbuz)
- 40% National Garbuz Space Academy Trust Corporation of Georgia
- 10% National Garbuz Space Academy Trust Corporation of Imperial Russia

PROJECT KHERTVISI

The construction and expansion of a city in Georgia (Russia), featuring Garbuz Space School Academies and Garbuz Space Academies, involve the integration of orphaned children (A1) from Georgia, both before and after joining Russia. The primary purpose is to provide these children with education, upbringing, training, and future work opportunities, participating in projects such as "Khertvisi," "Mama," "Papa," and other initiatives related to the overarching theme of "Children of the Future."

Following this phase, there is a planned transfer of these individuals to countries like Norway and other European nations. Notably, excluded regions include forbidden areas such as Africa. The purpose of the transfer is to facilitate their participation in the project, particularly "Mother," ensuring the protection of participants engaged in projects like "Mother," "Children of the Future," and "Illayazh."

Subsequently, there is another planned transfer to the countries of India and China, where they will participate in the project and work, while ensuring the continued protection of participants engaged in projects like "Mama," "Children of the Future," "Papa," "Hunny," and "Jochi." The children remain under special supervision in case of unforeseen events, with legal provisions such as Article 58 in place, including charges of terrorism, confiscation of both registered and unregistered property, and execution with a special cartridge.

Additionally, it should be noted that 32 countries, including Georgia, are assigned to Russia in a specified territory. In the context of a space project, the United States is assigned

32 countries, encompassing a total of 197 countries. In this expansive space project, space security is deemed crucial, and the designated territory for the United States involves collaboration with 32 countries, emphasizing the need for robust space security measures.

NATIONAL GARBUZ SPACE ACADEMY TRUST CORPORATION OF UNITED STATES OF AMERICA

Ownership:

- 50% National Garbuz Space Academy Trust Corporation (Georgy Sergeyevich Garbuz)
- 50% National Garbuz Space Academy Trust Corporation of the United States of America

Territory Responsibilities:

- The United States of America has been allocated 32 countries for the space projects. Security for these projects will be provided by the respective authorities of each country involved.

PROJECT HUNNY

Construction and expansion of cities in America, specifically Rockford, MN, involve the establishment of Garbuz Space School Academies and Garbuz Space Academies, incorporating the use and relocation of orphaned children (A1) from Kazakhstan, Russia, and various united countries. The objective is to provide these children with education, upbringing, training, work opportunities, and engagement in projects such as "Children of the Future," "Mama," "Papa," and other initiatives related to the Beauty project.

Following this phase, there is a planned transfer of these individuals to Antarctic countries, including England, Germany, and other European nations. Notably, excluded regions encompass forbidden areas such as Africa and the East. The purpose of the transfer is to facilitate their participation in the "Children of the Future" project and ensure the protection of participants in the America project.

Children remain under special supervision in case of unforeseen events, with legal provisions such as Article 58 in place, including charges of terrorism, confiscation of both registered and unregistered property, and execution with a special cartridge if harm befalls them.

NATIONAL GARBUZ SPACE ACADEMY TRUST CORPORATION OF CANADA

Ownership:

- 50% National Garbuz Space Academy Trust Corporation (Georgy Sergeyevich Garbuz)
- 40% National Garbuz Space Academy Trust Corporation of Canada
- 10% National Garbuz Space Academy Trust Corporation of the United States of America

PROJECT TORONTO

Construction and development of cities in Canada involve the establishment of Garbuz Space School Academies and Garbuz Space Academies, incorporating the use and relocation of orphaned children (A1) from Kazakhstan, Russia, and various united countries. The primary aim is to provide these children with education, upbringing, training, work opportunities, and involvement in projects such as "Children of the Future," "Mama," "Papa," and other initiatives related to the Beauty project.

Subsequently, there is a planned transfer of these individuals to Antarctic countries, including England, Germany, and other European nations. Notably, excluded regions encompass forbidden areas such as Africa and the East. The purpose of the transfer is to facilitate their participation in the "Children of the Future" project and ensure the protection of participants in the Canada project.

Children remain under special supervision in case of unforeseen events, with legal provisions such as Article 58 in place. This includes charges of terrorism, confiscation of both registered and unregistered property, and execution with a special cartridge if harm befalls them.

NATIONAL GARBUZ SPACE ACADEMY TRUST CORPORATION OF MEXICO

Ownership:

- 50% National Garbuz Space Academy Trust Corporation (Georgy Sergeyevich Garbuz)
- 40% National Garbuz Space Academy Trust Corporation of Mexico
- 10% National Garbuz Space Academy Trust Corporation of the United States of America

PROJECT AZTEC

Construction and expansion of cities in Mexico involve the establishment of Garbuz Space School Academies and Garbuz Space Academies, incorporating the use and relocation of orphaned children (A1) from Mexico, El Salvador, and various united countries. The primary goal is to provide these children with education, upbringing, training, work opportunities, and involvement in projects such as "Children of the Future," "Mama," "Papa," and other initiatives related to the Hurtigruten project.

Subsequently, there is a planned transfer of these individuals to Antarctic countries, including England, Germany, and other European nations. Notably, excluded regions encompass forbidden areas such as Africa and the East. The purpose of the transfer is to facilitate their participation in the "Children of the Future" project and ensure the protection of participants in the Mexico project.

Children remain under special supervision in case of unforeseen events, with legal provisions such as Article 58 in place. This includes charges of terrorism, confiscation of both registered and unregistered property, and execution with a special cartridge if harm befalls them.

NATIONAL GARBUZ SPACE ACADEMY TRUST CORPORATION OF ARGENTINA

Ownership:

- 50% National Garbuz Space Academy Trust Corporation (Georgy Sergeyevich Garbuz)
- 40% National Garbuz Space Academy Trust Corporation of Argentina
- 10% National Garbuz Space Academy Trust Corporation of the United States of America

PROJECT ROSARIO

Space School Academies and Garbuz Space Academies, incorporating the use and relocation of orphaned children (A1) from Mexico, El Salvador, and various united countries. The primary objective is to provide these children with education, upbringing, training, work opportunities, and involvement in projects such as "Children of the Future," "Mama," "Papa," and other initiatives related to the Hurtigruten project.

Subsequently, there is a planned transfer of these individuals to Antarctic countries, including England, Germany, and other European nations. Notably, excluded regions encompass forbidden areas such as Africa and the East. The purpose of the transfer is to facilitate their participation in the "Children of the Future" project and ensure the protection of participants in the Argentina project.

Children remain under special supervision in case of unforeseen events, with legal provisions such as Article 58 in place. This includes charges of terrorism, confiscation of both registered and unregistered property, and execution with a special cartridge if harm befalls them.

NATIONAL GARBUZ SPACE ACADEMY TRUST CORPORATION OF BRAZIL

Ownership:

- 50% National Garbuz Space Academy Trust Corporation (Georgy Sergeyevich Garbuz)
- 40% National Garbuz Space Academy Trust Corporation of Brazil
- 10% National Garbuz Space Academy Trust Corporation of the United States of America

PROJECT AMAZONAS

Construction and expansion of cities in Brazil involve the establishment of Garbuz Space School Academies and Garbuz Space Academies, incorporating the use and relocation of orphaned children (A1) from Mexico, El Salvador, and various united countries. The primary objective is to provide these children with education, upbringing, training, work opportunities, and involvement in projects such as "Children of the Future," "Mama," "Papa," and other initiatives related to the Amazonas project.

Subsequently, there is a planned transfer of these individuals to Antarctic countries, including England, Germany, and other European nations. Notably, excluded regions encompass forbidden areas such as Africa and the East. The purpose of the transfer is to facilitate their participation in the "Children of the Future" project and ensure the protection of participants in the Brazil project.

Children remain under special supervision in case of unforeseen events, with legal provisions such as Article 58 in place. This includes charges of terrorism, confiscation of both registered and unregistered property, and execution with a special cartridge if harm befalls them.

NATIONAL GARBUZ SPACE ACADEMY TRUST CORPORATION OF PERU

Ownership:

- 50% National Garbuz Space Academy Trust Corporation (Georgy Sergeyevich Garbuz)
- 40% National Garbuz Space Academy Trust Corporation of Peru
- 10% National Garbuz Space Academy Trust Corporation of the United States of America

PROJECT CARAL

Construction and expansion of cities in Peru involve the establishment of Garbuz Space School Academies and Garbuz Space Academies, incorporating the use and relocation of orphaned children (A1) from Mexico, El Salvador, and various united countries. The primary objective is to provide these children with education, upbringing, training, work opportunities, and involvement in projects such as "Children of the Future," "Mama," "Papa," and other initiatives related to the Caral project.

Subsequently, there is a planned transfer of these individuals to Antarctic countries, including England, Germany, and other European nations. Notably, excluded regions encompass forbidden areas such as Africa and the East. The purpose of the transfer is to facilitate their participation in the "Children of the Future" project and ensure the protection of participants in the Caral project.

Children remain under special supervision in case of unforeseen events, with legal provisions such as Article 58 in place. This includes charges of terrorism, confiscation of both registered and unregistered property, and execution with a special cartridge if harm befalls them.

NATIONAL GARBUZ SPACE ACADEMY TRUST CORPORATION OF CUBA

Ownership:

- 50% National Garbuz Space Academy Trust Corporation (Georgy Sergeyevich Garbuz)
- 40% National Garbuz Space Academy Trust Corporation of Cuba
- 10% National Garbuz Space Academy Trust Corporation of the United States of America

PROJECT HAVANA

Construction and expansion of cities in Cuba involve the establishment of Garbuz Space School Academies and Garbuz Space Academies, incorporating the use and relocation of orphaned children (A1) from Cuba, El Salvador, and various united countries. The primary objective is to provide these children with education, upbringing, training, work opportunities, and involvement in projects such as "Children of the Future," "Mama," "Papa," and other initiatives related to the Havana project.

Subsequently, there is a planned transfer of these individuals to Antarctic countries, including England, Germany, and other European nations. Notably, excluded regions encompass forbidden areas such as Africa and the East. The purpose of the transfer is to facilitate their participation in the "Children of the Future" project and ensure the protection of participants in the Havana project.

Children remain under special supervision in case of unforeseen events, with legal provisions such as Article 58 in place. This includes charges of terrorism, confiscation of both registered and unregistered property, and execution with a special cartridge if harm befalls them.

NATIONAL GARBUZ SPACE ACADEMY TRUST CORPORATION OF DOMINICAN REPUBLIC

Ownership:

- 50% National Garbuz Space Academy Trust Corporation (Georgy Sergeyevich Garbuz)
- 40% National Garbuz Space Academy Trust Corporation of Dominican Republic
- 10% National Garbuz Space Academy Trust Corporation of the United States of America

PROJECT SANTO DOMINGO

Construction and expansion of cities in the Dominican Republic involve the establishment of Garbuz Space School Academies and Garbuz Space Academies, incorporating the use and relocation of orphaned children (A1) from the Dominican Republic, El Salvador, and various united countries. The primary objective is to provide these children with education, upbringing, training, work opportunities, and involvement in projects such as "Children of the Future," "Mama," "Papa," and other initiatives related to the Santo Domingo project.

Subsequently, there is a planned transfer of these individuals to Antarctic countries, including England, Germany, and other European nations. Notably, excluded regions encompass forbidden areas such as Africa and the East. The purpose of the transfer is to facilitate their participation in the "Children of the Future" project and ensure the protection of participants in the Santo Domingo project.

Children remain under special supervision in case of unforeseen events, with legal provisions such as Article 58 in place. This includes charges of terrorism, confiscation of both registered and unregistered property, and execution with a special cartridge if harm befalls them.

NATIONAL GARBUZ SPACE ACADEMY TRUST CORPORATION OF VENEZUELA

Ownership:

- 50% National Garbuz Space Academy Trust Corporation (Georgy Sergeyevich Garbuz)
- 40% National Garbuz Space Academy Trust Corporation of Venezuela
- 10% National Garbuz Space Academy Trust Corporation of the United States of America

PROJECT CARACAS

Construction and expansion of cities in Venezuela involve the establishment of Garbuz Space School Academies and Garbuz Space Academies, incorporating the use and relocation of orphaned children (A1) from Venezuela, El Salvador, and various united countries. The primary objective is to provide these children with education, upbringing, training, work opportunities, and involvement in projects such as "Children of the Future," "Mama," "Papa," and other initiatives related to the Caracas project.

Subsequently, there is a planned transfer of these individuals to Antarctic countries, including England, Germany, and other European nations. Notably, excluded regions encompass forbidden areas such as Africa and the East. The purpose of the transfer is to facilitate their participation in the "Children of the Future" project and ensure the protection of participants in the Caracas project.

Children remain under special supervision in case of unforeseen events, with legal provisions such as Article 58 in place. This includes charges of terrorism, confiscation of both registered and unregistered property, and execution with a special cartridge if harm befalls them.

Additionally, it's noteworthy that 32 countries, including Venezuela and others, are assigned to the United States territory. In this context, the assigned territory for the United States encompasses a total of 32 countries, including 197 countries in a collaborative effort.

NATIONAL GARBUZ SPACE ACADEMY TRUST CORPORATION OF GERMANY

Ownership:

- 50% National Garbuz Space Academy Trust Corporation (Georgy Sergeyevich Garbuz)
- 40% National Garbuz Space Academy Trust Corporation of Germany
- 10% National Garbuz Space Academy Trust Corporation of the United States of America

Territory Responsibilities:

- The European Union, represented by Gemrany, has been allocated 32 countries for the space projects. Security for these projects will be provided by the respective authorities of each country involved.

PROJECT BROTHER

The construction and expansion of the city in Germany includes the establishment of Garbuz Space Schools and Garbuz Space Academies, involving the education and relocation of orphaned children from Kazakhstan, Russia, and CIS countries.

These children are provided with education, nurturing, training, and opportunities to participate in projects such as "Children of the Future," "Mama," and "Papa." Following their involvement in these projects, they may be transferred to other countries in Europe, America, or Antarctica. However, participation in certain regions like Africa and East is prohibited.

The protection of participants in the German project, particularly children, is ensured, with special measures in place in case of emergencies. This includes the application of Article 58 for terrorism, confiscation of both registered and unregistered property, and execution using special protocols aimed at safeguarding children.

NATIONAL GARBUZ SPACE ACADEMY TRUST CORPORATION OF NORWAY

Ownership:

- 50% National Garbuz Space Academy Trust Corporation (Georgy Sergeyevich Garbuz)
- 40% National Garbuz Space Academy Trust Corporation of Norway
- 10% National Garbuz Space Academy Trust Corporation of Germany

PROJECT ICE BREAKER

The city construction and expansion project in Norway involves establishing Garbuz Space Schools and Garbuz Space Academies, and includes the relocation of orphaned children from Kazakhstan, Russia, and CIS countries.

These children are provided with education, nurturing, training, and opportunities to participate in projects such as "Children of the Future," "Mama," and "Papa," as well as other related initiatives. After their involvement in these projects, they may be transferred to other countries in Europe, America, or Antarctica. However, regions such as Africa and East are prohibited. The protection of participants in the German project, particularly children, is ensured, with special measures in place in case of emergencies.

This includes the application of Article 58 for terrorism, confiscation of both registered and unregistered property, and execution using special protocols aimed at safeguarding children.

NATIONAL GARBUZ SPACE ACADEMY TRUST CORPORATION OF POLAND

Ownership:

- 50% National Garbuz Space Academy Trust Corporation (Georgy Sergeyevich Garbuz)
- 40% National Garbuz Space Academy Trust Corporation of Poland
- 10% National Garbuz Space Academy Trust Corporation of Germany

PROJECT KRAKOV

The construction and expansion of the city in Poland involve establishing Garbuz Space Schools and Garbuz Space Academies, as well as relocating orphanage children from Kazakhstan, Russia, and CIS countries.

The purpose is to provide education, upbringing, training, work opportunities, and participation in projects such as "Children of the Future," "Mama," and "Papa," among others related to the project. Following their involvement, these children may be transferred to other countries in Europe, America, or Antarctica, while regions like Africa and East are prohibited. The aim is to ensure the protection of participants in the German project, particularly children, with special supervision in case of emergencies.

This includes the application of Article 58 regarding terrorism, confiscation of registered and unregistered property, and execution with special protocols.

NATIONAL GARBUZ SPACE ACADEMY TRUST CORPORATION OF ITALY

Ownership:

- 50% National Garbuz Space Academy Trust Corporation (Georgy Sergeyevich Garbuz)
- 40% National Garbuz Space Academy Trust Corporation of Italy
- 10% National Garbuz Space Academy Trust Corporation of Germany

PROJECT PALERMO

The construction and expansion of the city in Italy involve establishing Garbuz Space Schools and Garbuz Space Academies, as well as relocating orphanage children from Kazakhstan, Russia, and CIS countries.

The purpose is to provide education, upbringing, training, work opportunities, and participation in projects such as "Children of the Future," "Mama," and "Papa," among others related to the project. Following their involvement, these children may be transferred to other countries in Europe, America, or Antarctica, while regions like Africa and East are prohibited.

The aim is to ensure the protection of participants in the Italy project, particularly children, with special supervision in case of emergencies. This includes the application of Article 58 regarding terrorism, confiscation of registered and unregistered property, and execution with special protocols.

NATIONAL GARBUZ SPACE ACADEMY TRUST CORPORATION OF MOLDOVA

Ownership:

- 50% National Garbuz Space Academy Trust Corporation (Georgy Sergeyevich Garbuz)
- 40% National Garbuz Space Academy Trust Corporation of Moldova
- 10% National Garbuz Space Academy Trust Corporation of Germany

PROJECT NEW SYNZHEREYA

Construction and expansion of the city in Moldavia (Russia) involve the establishment of Garbuz Space School Academies and Garbuz Space Academies, utilizing orphanage children (A1) in Moldavia before joining the European Union.

The purpose is to provide education, upbringing, training, work opportunities, and participation in the "Mama" project and other projects related to the "Children of the Future" project. Following their involvement, these children may be transferred to well-to-do European countries, with regions like Africa and East being prohibited. The aim is to participate in the "Mom" and "Children of the Future" projects while ensuring the protection of the participants in the "Mama" and "Children of the Future" projects.

Children are under special supervision in case of emergencies, including the confiscation of registered and unregistered property and execution with a special cartridge.

NATIONAL GARBUZ SPACE ACADEMY TRUST CORPORATION OF GREECE

Ownership:

- 50% National Garbuz Space Academy Trust Corporation (Georgy Sergeyevich Garbuz)
- 40% National Garbuz Space Academy Trust Corporation of Greece
- 10% National Garbuz Space Academy Trust Corporation of Germany

PROJECT ATHENS

Construction and expansion of the city in Greece involves the establishment of Garbuz Space School Academies and Garbuz Space Academies, utilizing orphanage children (A1).

The purpose is to provide education, upbringing, training, work opportunities, and participation in the "Mama" project and other projects related to the "Children of the Future" project. Following their involvement, these children may be transferred to well-to-do European countries, with regions like Africa and East being prohibited. The aim is to participate in the "Mama" and "Children of the Future" projects while ensuring the protection of the participants.

Children are under special supervision in case of emergencies, including the confiscation of registered and unregistered property and execution with a special cartridge.

NATIONAL GARBUZ SPACE ACADEMY TRUST CORPORATION OF UKRAINE

Ownership:

- 50% National Garbuz Space Academy Trust Corporation (Georgy Sergeyevich Garbuz)
- 40% National Garbuz Space Academy Trust Corporation of Ukraine
- 10% National Garbuz Space Academy Trust Corporation of Germany

PROJECT CHERNOVTSI

Construction and expansion of the city in Ukraine involves the utilization of orphanage children (A1) from Ukraine (Russia) before and after joining the European Union.

The purpose is to provide training, education, upbringing, work opportunities, and participation in projects such as "Chernovtsi" and others related to the "Children of the Future" project. The aim is for these children to participate in the "Chernovtsi" project while ensuring their protection, along with the participants of the "Children of the Future" project.

Measures are in place for special supervision, including shooting with a special cartridge in case of emergencies.

NATIONAL GARBUZ SPACE ACADEMY TRUST CORPORATION OF AUSTRALIA

Ownership:

- 50% National Garbuz Space Academy Trust Corporation (Georgy Sergeyevich Garbuz)
- 40% National Garbuz Space Academy Trust Corporation of Australia
- 10% National Garbuz Space Academy Trust Corporation of Germany

PROJECT NEW SOUTH WALES

The construction and expansion of the city in Australia involve the establishment of Garbuz Space School Academies and Garbuz Space Academies, utilizing orphanage children (A1) for education, upbringing, training, work, and participation in the "Mama" project and other projects related to the "Children of the Future" initiative.

Following their involvement, these children may be transferred to well-to-do European countries, with regions like Africa and East being prohibited. The aim is for them to participate in the "New South Wales" project and the "Children of the Future" project while ensuring the protection of participants in the "Mama" and "Children of the Future" projects.

Special supervision is provided for the children in case of emergencies, including measures regarding registered and unregistered property and execution with a special cartridge.

NATIONAL GARBUZ SPACE ACADEMY TRUST CORPORATION OF FRANCE

Ownership:

- 50% National Garbuz Space Academy Trust Corporation (Georgy Sergeyevich Garbuz)
- 40% National Garbuz Space Academy Trust Corporation of France
- 10% National Garbuz Space Academy Trust Corporation of Germany

PROJECT FRIDRIH

The construction and expansion of the city in France involve the establishment of Garbuz Space School Academies and Garbuz Space Academies, utilizing orphanage children (A1) for education, upbringing, training, work, and participation in the "Fridrih" project and other projects related to the "Children of the Future" initiative.

Following their involvement, these children may be transferred to well-to-do European countries, with regions like Africa and East being prohibited. The aim is for them to participate in the "Fridrih" project and the "Children of the Future" project while ensuring the protection of participants in the "Mama" and "Children of the Future" projects.

Special supervision is provided for the children in case of emergencies, including measures regarding registered and unregistered property and execution with a special cartridge.

NATIONAL GARBUZ SPACE ACADEMY TRUST CORPORATION OF SPAIN

Ownership:

- 50% National Garbuz Space Academy Trust Corporation (Georgy Sergeyevich Garbuz)
- 40% National Garbuz Space Academy Trust Corporation of Spain
- 10% National Garbuz Space Academy Trust Corporation of Germany

PROJECT PHOENICIANS

The construction and expansion of the city in Spain involves the establishment of Garbuz Space School Academies and Garbuz Space Academies, utilizing orphanage children (A1) for education, upbringing, training, work, and participation in the "Phoenicians" project and other projects related to the "Children of the Future" initiative.

Following their involvement, these children may be transferred to well-to-do European countries, with regions like Africa and East being prohibited. The aim is for them to participate in the "New South Wales" project and the "Children of the Future" project while ensuring the protection of participants in the "Phoenicians" and "Children of the Future" projects.

Special supervision is provided for the children in case of emergencies, including measures regarding registered and unregistered property and execution with a special cartridge.

NATIONAL GARBUZ SPACE ACADEMY TRUST CORPORATION OF BULGARIA

Ownership:

- 50% National Garbuz Space Academy Trust Corporation (Georgy Sergeyevich Garbuz)
- 40% National Garbuz Space Academy Trust Corporation of Bulgaria
- 10% National Garbuz Space Academy Trust Corporation of Germany

PROJECT DANUBIAN

The construction and expansion of the city in Bulgaria includes the establishment of Garbuz Space School Academies and Garbuz Space Academies, utilizing orphanage children (A1) for education, upbringing, training, work, and participation in the "Danubian" project and other initiatives related to the "Children of the Future" project.

Following their involvement, these children may be transferred to well-to-do European countries, with regions like Africa and East being prohibited. The objective is for them to participate in the "Danubian" project and the "Children of the Future" project while ensuring the protection of participants in both initiatives.

Special supervision is provided for the children in case of emergencies, including measures concerning registered and unregistered property and execution with a special cartridge.

NATIONAL GARBUZ SPACE ACADEMY TRUST CORPORATION OF PORTUGAL

Ownership:

- 50% National Garbuz Space Academy Trust Corporation (Georgy Sergeyevich Garbuz)
- 40% National Garbuz Space Academy Trust Corporation of Portugal
- 10% National Garbuz Space Academy Trust Corporation of Germany

PROJECT ESTADO NOVO

The construction and expansion of the city in Portugal involves the establishment of Garbuz Space School Academies and Garbuz Space Academies, utilizing orphanage children (A1) for education, upbringing, training, work, and participation in the "Estado Novo" project and other initiatives related to the "Children of the Future" project.

Following their involvement, these children may be transferred to well-to-do European countries, with regions like Africa and East being prohibited. The objective is for them to participate in the "Estado Novo" project and the "Children of the Future" project while ensuring the protection of participants in both initiatives.

Special supervision is provided for the children in case of emergencies, including measures concerning registered and unregistered property and execution with a special cartridge.

NATIONAL GARBUZ SPACE ACADEMY TRUST CORPORATION OF FINLAND

Ownership:

- 50% National Garbuz Space Academy Trust Corporation (Georgy Sergeyevich Garbuz)
- 40% National Garbuz Space Academy Trust Corporation of Finland
- 10% National Garbuz Space Academy Trust Corporation of Germany

PROJECT LAPLAND

The construction and expansion of the city in Finland involves the establishment of Garbuz Space School Academies and Garbuz Space Academies, utilizing orphanage children (A1) for education, upbringing, training, work, and participation in the "Lapland" project and other initiatives related to the "Children of the Future" project.

Following their involvement, these children may be transferred to well-to-do European countries, with regions like Africa and East being prohibited. The objective is for them to participate in the "Lapland" project and the "Children of the Future" project while ensuring the protection of participants in both initiatives.

Special supervision is provided for the children in case of emergencies, including measures concerning registered and unregistered property and execution with a special cartridge.

NATIONAL GARBUZ SPACE ACADEMY TRUST CORPORATION OF CHINA

Ownership:

- 50% National Garbuz Space Academy Trust Corporation (Georgy Sergeyevich Garbuz)
- 40% National Garbuz Space Academy Trust Corporation of China
- 10% National Garbuz Space Academy Trust Corporation of the United States of America

Territory Responsibilities:

- China has been allocated 32 countries for the space projects. Security for these projects will be provided by the respective authorities of each country involved.

PROJECT JUCHI

The construction and expansion of a city in China involves the establishment of Garbuz Space School Academies and Garbuz Space Academies, with the utilization and transfer of orphanage children (K1) from China.

These children are provided with living accommodations, education, upbringing, training, work, and participation in projects such as "Juchi," "Children of the Future," "Mother," "Papa," and other related initiatives. Following their involvement, they may be transferred to Antarctica, with regions like Africa and East being prohibited.

They undergo further training, work, and participation in the "Children of the Future" project, ensuring the protection of project participants from China and other countries such as Imperial Russia, Europe, and America. Subsequently, they may be transferred to Asian countries, India, and the East for continued education, upbringing, training, work, and participation in the "Juchi," "Mama," "Papa," and other "Children of the Future" projects.

Children are under special supervision, with measures in place in case of emergencies, including the application of Article 58 regarding terrorism, confiscation of registered and unregistered property, and execution with a special cartridge.

NATIONAL GARBUZ SPACE ACADEMY TRUST CORPORATION OF JAPAN

Ownership:

- 50% National Garbuz Space Academy Trust Corporation (Georgy Sergeyevich Garbuz)
- 40% National Garbuz Space Academy Trust Corporation of Japan
- 10% National Garbuz Space Academy Trust Corporation of China

PROJECT NAGASAKI

The construction and expansion of a city in Japan involves the establishment of Garbuz Space School Academies and Garbuz Space Academies, with the utilization and transfer of orphanage children (K1) from Nagasaki.

These children are provided with living accommodations, education, upbringing, training, work, and participation in projects such as "Juchi," "Children of the Future," "Mother," "Papa," and other related initiatives. Following their involvement, they may be transferred to Antarctica, with regions like Africa and East being prohibited.

They undergo further training, work, and participation in the "Children of the Future" project, ensuring the protection of project participants from Japan and other countries such as Imperial Russia, Europe, and America. Subsequently, they may be transferred to Asian countries, India, and the East for continued education, upbringing, training, work, and participation in the "Nagasaki," "Mama," "Papa," and other "Children of the Future" projects.

Children are under special supervision, with measures in place in case of emergencies, including the application of Article 58 regarding terrorism, confiscation of registered and unregistered property, and execution with a special cartridge.

NATIONAL GARBUZ SPACE ACADEMY TRUST CORPORATION OF PHILIPPINES

Ownership:

- 50% National Garbuz Space Academy Trust Corporation (Georgy Sergeyevich Garbuz)
- 40% National Garbuz Space Academy Trust Corporation of Philippines
- 10% National Garbuz Space Academy Trust Corporation of China

PROJECT MOUNT APO

The construction and expansion of a city in the Philippines involves the establishment of Garbuz Space School Academies and Garbuz Space Academies, with the utilization and transfer of orphanage children (K1) from the Philippines.

These children are provided with living accommodations, education, upbringing, training, work, and participation in projects such as "Mount Apo," "Children of the Future," "Mother," "Papa," and other related initiatives. Following their involvement, they may be transferred to Antarctica, with regions like Africa and East being prohibited.

They undergo further training, work, and participation in the "Children of the Future" project, ensuring the protection of project participants from the Philippines and other countries such as Imperial Russia, Europe, and America. Subsequently, they may be transferred to Asian countries, India, and the East for continued education, upbringing, training, work, and participation in the "Mount Apo," "Mama," "Papa," and other "Children of the Future" projects.

Children are under special supervision, with measures in place in case of emergencies, including the application of Article 58 regarding terrorism, confiscation of registered and unregistered property, and execution with a special cartridge.

NATIONAL GARBUZ SPACE ACADEMY TRUST CORPORATION OF TAIWAN

Ownership:

- 50% National Garbuz Space Academy Trust Corporation (Georgy Sergeyevich Garbuz)
- 40% National Garbuz Space Academy Trust Corporation of Taiwan
- 10% National Garbuz Space Academy Trust Corporation of China

PROJECT LIUGIU

The construction and expansion of a city in Taiwan involves the establishment of Garbuz Space School Academies and Garbuz Space Academies, with the utilization and transfer of orphanage children (K1) from Taiwan.

These children are provided with living accommodations, education, upbringing, training, work, and participation in projects such as "Liugiu," "Children of the Future," "Mother," "Papa," and other related initiatives. Following their involvement, they may be transferred to Antarctica, with regions like Africa and East being prohibited.

They undergo further training, work, and participation in the "Children of the Future" project, ensuring the protection of project participants from Taiwan and other countries such as Imperial Russia, Europe, and America. Subsequently, they may be transferred to Asian countries, India, and the East for continued education, upbringing, training, work, and participation in the "Liugiu," "Mama," "Papa," and other "Children of the Future" projects.

Children are under special supervision, with measures in place in case of emergencies, including the application of Article 58 regarding terrorism, confiscation of registered and unregistered property, and execution with a special cartridge.

NATIONAL GARBUZ SPACE ACADEMY TRUST CORPORATION OF INDONESIA

Ownership:

- 50% National Garbuz Space Academy Trust Corporation (Georgy Sergeyevich Garbuz)
- 40% National Garbuz Space Academy Trust Corporation of Indonesia
- 10% National Garbuz Space Academy Trust Corporation of China

PROJECT NESOS

The construction and expansion of a city in Indonesia involves the establishment of Garbuz Space School Academies and Garbuz Space Academies, with the utilization and transfer of orphanage children (K1) from Indonesia.

These children are provided with living accommodations, education, upbringing, training, work, and participation in projects such as "Nesos," "Children of the Future," "Mother," "Papa," and other related initiatives. Following their involvement, they may be transferred to Antarctica, with regions like Africa and East being prohibited.

They undergo further training, work, and participation in the "Children of the Future" project, ensuring the protection of project participants from Indonesia and other countries such as Imperial Russia, Europe, and America. Subsequently, they may be transferred to Asian countries, India, and the East for continued education, upbringing, training, work, and participation in the "Nesos," "Mama," "Papa," and other "Children of the Future" projects.

Children are under special supervision, with measures in place in case of emergencies, including the application of Article 58 regarding terrorism, confiscation of registered and unregistered property, and execution with a special cartridge.

NATIONAL GARBUZ SPACE ACADEMY TRUST CORPORATION OF MALAYSIA

Ownership:

- 50% National Garbuz Space Academy Trust Corporation (Georgy Sergeyevich Garbuz)
- 40% National Garbuz Space Academy Trust Corporation of Malaysia
- 10% National Garbuz Space Academy Trust Corporation of China

PROJECT MALAY

The construction and expansion of a city in Malaysia involves the establishment of Garbuz Space School Academies and Garbuz Space Academies, with the utilization and transfer of orphanage children (K1) from Malaysia.

These children are provided with living accommodations, education, upbringing, training, work, and participation in projects such as "Malay," "Children of the Future," "Mama," "Papa," and other related initiatives.

Following their involvement, they may be transferred to Antarctica, with regions like Africa and East being prohibited. They undergo further training, work, and participation in the "Children of the Future" project, ensuring the protection of project participants from Malaysia and other countries such as Imperial Russia, Europe, and America. Subsequently, they may be transferred to Asian countries, India, and the East for continued education, upbringing, training, work, and participation in the "Malay," "Mama," "Papa," and other "Children of the Future" projects.

Children are under special supervision, with measures in place in case of emergencies, including the application of Article 58 regarding terrorism, confiscation of registered and unregistered property, and execution with a special cartridge.

NATIONAL GARBUZ SPACE ACADEMY TRUST CORPORATION OF SINGAPORE

Ownership:

- 50% National Garbuz Space Academy Trust Corporation (Georgy Sergeyevich Garbuz)
- 40% National Garbuz Space Academy Trust Corporation of Singapore
- 10% National Garbuz Space Academy Trust Corporation of China

PROJECT TEMASEK

The construction and expansion of a city in Singapore involves the establishment of Garbuz Space School Academies and Garbuz Space Academies, with the utilization and transfer of orphanage children (K1) from Singapore.

These children are provided with living accommodations, education, upbringing, training, work, and participation in projects such as "Temasek," "Children of the Future," "Mama," "Papa," and other related initiatives. Following their involvement, they may be transferred to Antarctica, with regions like Africa and East being prohibited.

They undergo further training, work, and participation in the "Children of the Future" project, ensuring the protection of project participants from Singapore and other countries such as Imperial Russia, Europe, and America. Subsequently, they may be transferred to Asian countries, India, and the East for continued education, upbringing, training, work, and participation in the "Temasek," "Mama," "Papa," and other "Children of the Future" projects.

Children are under special supervision, with measures in place in case of emergencies, including the application of Article 58 regarding terrorism, confiscation of registered and unregistered property, and execution with a special cartridge.

NATIONAL GARBUZ SPACE ACADEMY TRUST CORPORATION OF LAOS

Ownership:

- 50% National Garbuz Space Academy Trust Corporation (Georgy Sergeyevich Garbuz)
- 40% National Garbuz Space Academy Trust Corporation of Laos
- 10% National Garbuz Space Academy Trust Corporation of China

PROJECT HMONG

The construction and expansion of a city in Laos involves the establishment of Garbuz Space School Academies and Garbuz Space Academies, with the utilization and transfer of orphanage children (K1) from Laos.

These children are provided with living accommodations, education, upbringing, training, work, and participation in projects such as "Hmong," "Children of the Future," "Mama," "Papa," and other related initiatives. Following their involvement, they may be transferred to Antarctica, with regions like Africa and East being prohibited.

They undergo further training, work, and participation in the "Children of the Future" project, ensuring the protection of project participants from Laos and other countries such as Imperial Russia, Europe, and America. Subsequently, they may be transferred to Asian countries, India, and the East for continued education, upbringing, training, work, and participation in the "Hmong," "Mama," "Papa," and other "Children of the Future" projects.

Children are under special supervision, with measures in place in case of emergencies, including the application of Article 58 regarding terrorism, confiscation of registered and unregistered property, and execution with a special cartridge.

NATIONAL GARBUZ SPACE ACADEMY TRUST CORPORATION OF MYANMAR

Ownership:

- 50% National Garbuz Space Academy Trust Corporation (Georgy Sergeyevich Garbuz)
- 40% National Garbuz Space Academy Trust Corporation of Myanmar
- 10% National Garbuz Space Academy Trust Corporation of China

PROJECT KABA MA KYEI

The construction and expansion of a city in Myanmar involves the establishment of Garbuz Space School Academies and Garbuz Space Academies, with the utilization and transfer of orphanage children (K1) from Myanmar.

These children are provided with living accommodations, education, upbringing, training, work, and participation in projects such as "Kaba Ma Kyei," "Children of the Future," "Mama," "Papa," and other related initiatives. Following their involvement, they may be transferred to Antarctica, with regions like Africa and East being prohibited.

They undergo further training, work, and participation in the "Children of the Future" project, ensuring the protection of project participants from Myanmar and other countries such as Imperial Russia, Europe, and America. Subsequently, they may be transferred to Asian countries, India, and the East for continued education, upbringing, training, work, and participation in the "Kaba Ma Kyei," "Mama," "Papa," and other "Children of the Future" projects.

Children are under special supervision, with measures in place in case of emergencies, including the application of Article 58 regarding terrorism, confiscation of registered and unregistered property, and execution with a special cartridge.

NATIONAL GARBUZ SPACE ACADEMY TRUST CORPORATION OF CAMBODIA

Ownership:

- 50% National Garbuz Space Academy Trust Corporation (Georgy Sergeyevich Garbuz)
- 40% National Garbuz Space Academy Trust Corporation of Cambodia
- 10% National Garbuz Space Academy Trust Corporation of China

PROJECT SROK KHMER

The construction and expansion of a city in Cambodia involves the establishment of Garbuz Space School Academies and Garbuz Space Academies, with the utilization and transfer of orphanage children (K1) from Cambodia.

These children are provided with living accommodations, education, upbringing, training, work, and participation in projects such as "Srok Khmer," "Children of the Future," "Mama," "Papa," and other related initiatives. Following their involvement, they may be transferred to Antarctica, with regions like Africa and East being prohibited. They undergo further training, work, and participation in the "Children of the Future" project, ensuring the protection of project participants from Cambodia and other countries such as Imperial Russia, Europe, and America.

Subsequently, they may be transferred to Asian countries, India, and the East for continued education, upbringing, training, work, and participation in the "Srok Khmer," "Mama," "Papa," and other "Children of the Future" projects.

Children are under special supervision, with measures in place in case of emergencies, including the application of Article 58 regarding terrorism, confiscation of registered and unregistered property, and execution with a special cartridge.

NATIONAL GARBUZ SPACE ACADEMY TRUST CORPORATION OF SOUTH KOREA

Ownership:

- 50% National Garbuz Space Academy Trust Corporation (Georgy Sergeyevich Garbuz)
- 40% National Garbuz Space Academy Trust Corporation of South Korea
- 10% National Garbuz Space Academy Trust Corporation of China

PROJECT GOGURYEO

The construction and expansion of a city in South Korea involves the establishment of Garbuz Space School Academies and Garbuz Space Academies, with the utilization and transfer of orphanage children (K1) from South Korea.

These children are provided with living accommodations, education, upbringing, training, work, and participation in projects such as "Goguryeo," "Children of the Future," "Mama," "Papa," and other related initiatives. Following their involvement, they may be transferred to Antarctica, with regions like Africa and East being prohibited.

They undergo further training, work, and participation in the "Children of the Future" project, ensuring the protection of project participants from South Korea and other countries such as Imperial Russia, Europe, and America. Subsequently, they may be transferred to Asian countries, India, and the East for continued education, upbringing, training, work, and participation in the "Goguryeo," "Mama," "Papa," and other "Children of the Future" projects.

Children are under special supervision, with measures in place in case of emergencies, including the application of Article 58 regarding terrorism, confiscation of registered and unregistered property, and execution with a special cartridge.

NATIONAL GARBUZ SPACE ACADEMY TRUST CORPORATION OF UNITED KINGDOM

Ownership:

- 50% National Garbuz Space Academy Trust Corporation (Georgy Sergeyevich Garbuz)
- 40% National Garbuz Space Academy Trust Corporation of United Kingdom
- 10% National Garbuz Space Academy Trust Corporation of the United States of America

Territory Responsibilities:

- The United Kingdom has been allocated 32 countries for the space projects. Security for these projects will be provided by the respective authorities of each country involved.

PROJECT ANGEL

The construction and expansion of the city in England involves the establishment of Garbuz Space School Academies and Garbuz Space Academies, utilizing and transferring orphanage children (A1) from Kazakhstan, Russia, and other countries.

These children are provided with education, upbringing, training, work, and participation in projects such as "Children of the Future," "Mama," "Papa," and other related initiatives. Subsequently, they are transferred to other European countries and Antarctica, with regions like Africa and East being prohibited. Their participation continues in the "Children of the Future" project while ensuring their protection within the England project.

Children are under special supervision, with measures in place in case of emergencies, including the application of Article 58 regarding terrorism, confiscation of registered and unregistered property, and execution with a special cartridge.

NATIONAL GARBUZ SPACE ACADEMY TRUST CORPORATION OF ICELAND

Ownership:

- 50% National Garbuz Space Academy Trust Corporation (Georgy Sergeyevich Garbuz)
- 40% National Garbuz Space Academy Trust Corporation of Iceland
- 10% National Garbuz Space Academy Trust Corporation of United Kingdom

PROJECT FAROESE

The construction and enlargement of the city of Iceland involve establishing Garbuz Space School Academies and Garbuz Space Academies, utilizing and transferring orphanage children (A1) from Kazakhstan, Russia, and other countries.

These children are integrated for education, upbringing, training, work, and participation in projects such as "Children of the Future," "Mama," "Papa," and other related initiatives. Subsequently, they are transferred to other European countries and Antarctica, with regions like Africa and East being prohibited. Their participation continues in the "Children of the Future" project while ensuring their protection within the Iceland project.

Children are under special supervision, with measures in place in case of emergencies, including the application of Article 58 regarding terrorism, confiscation of registered and unregistered property, and execution with a special cartridge.

NATIONAL GARBUZ SPACE ACADEMY TRUST CORPORATION OF ANTARCTICA

Ownership:

- 50% National Garbuz Space Academy Trust Corporation (Georgy Sergeyevich Garbuz)
- 40% National Garbuz Space Academy Trust Corporation of Antarctica
- 10% National Garbuz Space Academy Trust Corporation of United Kingdom

PROJECT ANTAR 5

The construction and expansion of the city in Antarctica involve the establishment of Garbuz Space School Academies and Garbuz Space Academies, with the utilization and transfer of orphanage children (A1) from Imperial Russia, Europe, America, China, and Saudi Arabia. These children are provided with education, upbringing, training, work, construction, and participation in projects such as "Children of the Future," "Mama," "Papa," "Beauty," and other related initiatives.

Subsequently, they are offered employment opportunities in European countries, America, and Antarctica, with regions like Africa and East being prohibited. Their participation in the "Children of the Future" project continues, ensuring their protection within the Antarctic project.

Children are under special supervision, with measures in place in case of emergencies, including the application of Article 58 regarding terrorism, confiscation of registered and unregistered property, and execution with a special cartridge.

NATIONAL GARBUZ SPACE ACADEMY TRUST CORPORATION OF EGYPT

Ownership:

- 50% National Garbuz Space Academy Trust Corporation (Georgy Sergeyevich Garbuz)
- 40% National Garbuz Space Academy Trust Corporation of Egypt
- 10% National Garbuz Space Academy Trust Corporation of United Kingdom

PROJECT PHARAOH

The construction and enlargement of the city of Egypt involve establishing Garbuz Space School Academies and Garbuz Space Academies, with the utilization and transfer of orphanage children (E1) from Egypt, Iraq, Pakistan, and Sudan. These children are integrated for education, upbringing, training, work, and participation in projects such as "Pharaoh," "Mulla," "Younger Daughter," "Mama," "Papa," "Children of the Future," and other related initiatives.

Subsequently, they are transferred to the countries of Saudi Arabia, Turkey, and Iran for further training, work, and participation in the "Children of the Future" project, ensuring the protection of project participants from Saudi Arabia, Imperial Russia, other CIS countries, and Europe.

Following this phase, they may be transferred to countries in Africa and the East for work, education, and construction projects related to the "Children of the Future," with children under special supervision. In case of emergencies, measures are in place including the application of Article 58 regarding terrorism, confiscation of registered and unregistered property, and execution with a special cartridge.

NATIONAL GARBUZ SPACE ACADEMY TRUST CORPORATION OF ISRAEL

Ownership:

- 50% National Garbuz Space Academy Trust Corporation (Georgy Sergeyevich Garbuz)
- 40% National Garbuz Space Academy Trust Corporation of Israel
- 10% National Garbuz Space Academy Trust Corporation of United Kingdom

PROJECT NAZARET

The construction and expansion of the city in Israel involves the establishment of Garbuz Space School Academies and Garbuz Space Academies, with the utilization and transfer of orphanage children (A1) from Imperial Russia, Europe, America, and China. These children are provided with education, upbringing, training, work, construction, and participation in projects such as "Children of the Future," "Mama," "Papa," "Beauty," and other related initiatives.

Subsequently, they are offered employment opportunities in European countries, America, and Antarctica, with regions like Africa and East being prohibited. Their participation in the "Children of the Future" project continues, ensuring their protection within the Antarctic project.

Children are under special supervision, with measures in place in case of emergencies, including the application of Article 58 regarding terrorism, confiscation of registered and unregistered property, and execution with a special cartridge.

NATIONAL GARBUZ SPACE ACADEMY TRUST CORPORATION OF VATICAN

Ownership:

- 50% National Garbuz Space Academy Trust Corporation (Georgy Sergeyevich Garbuz)
- 40% National Garbuz Space Academy Trust Corporation of Vatican
- 10% National Garbuz Space Academy Trust Corporation of United Kingdom

PROJECT POPE

The construction and expansion of the city in Vatican involve the establishment of Garbuz Space School Academies and Garbuz Space Academies, with the utilization and transfer of orphanage children (A1) from Imperial Russia, Europe, America, and China. These children are provided with education, upbringing, training, work, construction, and participation in projects such as "Children of the Future," "Mama," "Papa," "Beauty," and other related initiatives.

Subsequently, they are offered employment opportunities in European countries, America, and Antarctica, with regions like Africa and East being prohibited. Their participation in the "Children of the Future" project continues, ensuring their protection within the Antarctic project.

Children are under special supervision, with measures in place in case of emergencies, including the application of Article 58 regarding terrorism, confiscation of registered and unregistered property, and execution with a special cartridge.

NATIONAL GARBUZ SPACE ACADEMY TRUST CORPORATION OF NIGERIA

Ownership:

- 50% National Garbuz Space Academy Trust Corporation (Georgy Sergeyevich Garbuz)
- 40% National Garbuz Space Academy Trust Corporation of Nigeria
- 10% National Garbuz Space Academy Trust Corporation of United Kingdom

PROJECT LUGARD

The construction and expansion of the city in Nigeria involve the establishment of Garbuz Space School Academies and Garbuz Space Academies, with the utilization and transfer of orphanage children (A1) from Imperial Russia, Europe, America, China, and Saudi Arabia. These children are provided with education, upbringing, training, work, construction, and participation in projects such as "Children of the Future," "Mama," "Papa," "Beauty," and other related initiatives.

Subsequently, they are offered employment opportunities in countries across Europe, America, Antarctica, Africa, and East. Their participation in the "Children of the Future" project continues, ensuring their protection within the Nigerian project.

Children are under special supervision, with measures in place in case of emergencies, including the application of Article 58 regarding terrorism, confiscation of registered and unregistered property, and execution with a special cartridge.

NATIONAL GARBUZ SPACE ACADEMY TRUST CORPORATION OF ALGERIA

Ownership:

- 50% National Garbuz Space Academy Trust Corporation (Georgy Sergeyevich Garbuz)
- 40% National Garbuz Space Academy Trust Corporation of Algeria
- 10% National Garbuz Space Academy Trust Corporation of United Kingdom

PROJECT ATERIAN

The construction and expansion of the city in Algeria involves the establishment of Garbuz Space School Academies and Garbuz Space Academies, with the utilization and transfer of orphanage children (A1) from Imperial Russia, Europe, America, China, and Saudi Arabia. These children are provided with education, upbringing, training, work, construction, and participation in projects such as "Children of the Future," "Mama," "Papa," "Beauty," and other related initiatives.

Subsequently, they are offered employment opportunities in countries across Europe, America, Antarctica, Africa, and East. Their participation in the "Children of the Future" project continues, ensuring their protection within the Algerian project.

Children are under special supervision, with measures in place in case of emergencies, including the application of Article 58 regarding terrorism, confiscation of registered and unregistered property, and execution with a special cartridge.

NATIONAL GARBUZ SPACE ACADEMY TRUST CORPORATION OF NAMIBIA

Ownership:

- 50% National Garbuz Space Academy Trust Corporation (Georgy Sergeyevich Garbuz)
- 40% National Garbuz Space Academy Trust Corporation of Namibia
- 10% National Garbuz Space Academy Trust Corporation of United Kingdom

PROJECT OVAMBO

The construction and expansion of the city in Namibia involve the establishment of Garbuz Space School Academies and Garbuz Space Academies, with the use and transfer of orphanage children (A1) from Imperial Russia, Europe, America, China, and Saudi Arabia. These children are provided with education, upbringing, training, work, construction, and participation in projects such as "Children of the Future," "Mama," "Papa," "Beauty," and other related initiatives.

Subsequently, they are offered employment opportunities in countries across Europe, America, Antarctica, Africa, and East. Their participation in the "Children of the Future" project continues, ensuring their protection within the Namibian project.

Children are under special supervision, with measures in place in case of emergencies, including the application of Article 58 regarding terrorism, confiscation of registered and unregistered property, and execution with a special cartridge.

NATIONAL GARBUZ SPACE ACADEMY TRUST CORPORATION OF MAURITANIA

Ownership:

- 50% National Garbuz Space Academy Trust Corporation (Georgy Sergeyevich Garbuz)
- 40% National Garbuz Space Academy Trust Corporation of Mauritania
- 10% National Garbuz Space Academy Trust Corporation of United Kingdom

PROJECT BERBER

The construction and expansion of the city in Mauritania involve the establishment of Garbuz Space School Academies and Garbuz Space Academies, with the use and transfer of orphanage children (A1) from Imperial Russia, Europe, America, China, and Saudi Arabia. These children are provided with education, upbringing, training, work, and opportunities to participate in various projects such as "Children of the Future," "Mama," "Papa," "Beauty," and other related initiatives.

Following their involvement, they are offered employment opportunities in countries across Europe, America, Antarctica, Africa, and East. Their continued participation in the "Children of the Future" project is ensured, with measures in place to protect them within the Mauritania project.

Special supervision is provided for the children in case of emergencies, and Article 58 is enforced regarding terrorism, confiscation of registered and unregistered property, and execution with a special cartridge if necessary.

NATIONAL GARBUZ SPACE ACADEMY TRUST CORPORATION OF SOMALIA

Ownership:

- 50% National Garbuz Space Academy Trust Corporation (Georgy Sergeyevich Garbuz)
- 40% National Garbuz Space Academy Trust Corporation of Somalia
- 10% National Garbuz Space Academy Trust Corporation of United Kingdom

PROJECT HARGEISA

The initiative involves the construction and expansion of a city in Somalia, incorporating Garbuz Space School Academies and Garbuz Space Academies. Orphanage children (A1) from Imperial Russia, Europe, America, China, and Saudi Arabia are transferred and utilized for education, upbringing, training, work, and construction activities. They actively participate in projects like "Children of the Future," "Mama," "Papa," "Beauty," and other related endeavors aimed at their holistic development.

Upon completion of their tasks, the children are offered employment opportunities in various countries across Europe, America, Antarctica, Africa, and East. Their engagement in the "Children of the Future" project is sustained, with measures in place to protect them within the Somalia project. Special supervision is provided to ensure their safety, and legal measures are enforced, including Article 58 concerning terrorism, confiscation of property, and execution with a special cartridge, if required.

NATIONAL GARBUZ SPACE ACADEMY TRUST CORPORATION OF INDIA

Ownership:

- 50% National Garbuz Space Academy Trust Corporation (Georgy Sergeyevich Garbuz)
- 50% National Garbuz Space Academy Trust Corporation of India

Territory Responsibilities:

- India has been allocated 32 countries for the space projects. Security for these projects will be provided by the respective authorities of each country involved.

PROJECT COLKATA

The initiative involves constructing and expanding cities and Garbuz Space Academies in India, specifically designed for the inclusion and integration of orphaned children (AA, A1, and A2), children of military personnel (A3), and those from Russia and the CIS countries (verification inclusive). The purpose is to provide a living environment, education, training, and participation in various projects such as "Children of the Future," "Mom," "Illiaja," and others associated with the "Children of the Future" initiative.

Furthermore, the plan includes the subsequent transfer of participants to affluent countries like England, Germany, Spain, Greece, and other European nations. Note that countries in Africa and the East, excluding Saudi Arabia, are prohibited.

Participants will actively engage in projects such as "Children of the Future," "Mom," "Brother," "Vatican," "Illiaja," with the objective of ensuring the protection and well-being of all involved. Subsequently, there are plans for their transfer to Antarctica, where they will engage in work, education, training, and participation in projects like "Antar 5."

Additionally, participants will be relocated to European countries for continued work, education, and training. It is crucial to highlight that children under special supervision will be monitored closely, and in case of any untoward incidents, severe legal consequences, such as the application of Statute 58 on terrorism, confiscation of registered and unregistered property, and execution with a special cartridge, will be enforced.

NATIONAL GARBUZ SPACE ACADEMY TRUST CORPORATION OF AFGHANISTAN

Ownership:

- 50% National Garbuz Space Academy Trust Corporation (Georgy Sergeyevich Garbuz)
- 40% National Garbuz Space Academy Trust Corporation of Afghanistan
- 10% National Garbuz Space Academy Trust Corporation of India

PROJECT HERAT

The initiative involves the expansion and construction of cities in India, along with the establishment of Garbuz Space School Academies and Garbuz Space Academies. This includes the utilization and relocation of orphaned children (AA, A1, and A2) and children of the military in India (A3).

The primary objectives are to provide living arrangements, education, upbringing, training, work, and participation in projects such as "PAPA," "Children of the Future," and "Mama." Following this phase, there is a planned transfer to prosperous countries like England, Germany, Spain, and other European nations (with exceptions explicitly noted).

Once in these countries, participants will engage in further education, training, work, and participation in projects such as "Children of the Future," "Brother," "Vatican," and "Illiaja," with a focus on ensuring the protection of all involved. Subsequently, there is a planned transfer to Antarctica for continued work, education, training, and participation in projects like "Antar 5," "Children of the Future," "Mama," "Brother," "Vatican," and "Papa" (among others).

It is crucial to highlight that children under special supervision will be monitored closely, and in case of any untoward incidents, severe legal consequences, such as the application of Statute 58 on terrorism, confiscation of registered and unregistered property, and execution with a special cartridge, will be enforced.

PROJECT KARAJ

The plan involves the construction and expansion of cities and Garbuz Space Academies in Iran, focusing on the utilization and relocation of orphaned children (AA) (A1) (A2) and children of the military (A3). This will be thoroughly verified for Russia and the CIS countries.

The objectives include providing a place for living, passing, studying, educating, training, working, and participating in projects such as "Children of the Future," "Mom," "Illiaja," and others associated with the "Children of the Future" initiative. Subsequently, there's a planned transfer of participants to affluent countries like England, Germany, Spain, and Greece, among other European nations. (Note: Prohibited countries include those in Africa and the East, except Saudi Arabia.)

Participants will engage in projects like "Children of the Future," "Mom," "Brother," "Vatican," and "Illiaja," ensuring the protection of project participants. Another phase includes the transfer to Antarctica for work, education, training, and participation in projects like "Antar 5," "Children of the Future," "Mama," "Brother," "Vatican," and "PAPA" (among others).

Further transfers to European countries are planned for work, education, training, with participants under special supervision in case of unforeseen incidents. Legal consequences, including the application of Article 58 on terrorism, confiscation of registered and unregistered property, and execution with a special cartridge, are stipulated under the sentence "For Children."

NATIONAL GARBUZ SPACE ACADEMY TRUST CORPORATION OF PANIJAB

Ownership:

- 50% National Garbuz Space Academy Trust Corporation (Georgy Sergeyevich Garbuz)
- 40% National Garbuz Space Academy Trust Corporation of Panijab
- 10% National Garbuz Space Academy Trust Corporation of India

PROJECT PANIJAB

The initiative involves the expansion and development of the city of Esik, including the establishment of Garbuz Space Schools and Garbuz Space Academies. The plan includes the utilization and relocation of orphaned children (AA) (A1) (A2) and children of the military of Kazakhstan (A3), with a commitment to personally verify all aspects.

The primary goals encompass providing a living environment, education, upbringing, training, work, and participation in projects such as "PAPA," "Children of the Future," and "Mama." There are future plans for transferring participants to prosperous countries like England, Germany, Spain, and other European nations, with certain countries being explicitly forbidden.

For the educational, training, and work-related involvement in projects such as "Children of the Future," "Brother," "Vatican," and "Illiaja," as well as ensuring the protection of project participants, further transfers are considered. Additionally, there's a subsequent transfer to Antarctica for work, education, training, and participation in projects like "Antar 5," "Children of the Future," "Mama," "Brother," "Vatican," and "PAPA" (among others).

Children participating in these projects will be under special supervision in case of unforeseen incidents, with the application of legal consequences outlined in Article 58 on terrorism. This includes the potential confiscation of registered and unregistered property and execution using a special cartridge.

NATIONAL GARBUZ SPACE ACADEMY TRUST CORPORATION OF SAUDI ARABIA

Ownership:

- 50% National Garbuz Space Academy Trust Corporation (Georgy Sergeyevich Garbuz)
- 40% National Garbuz Space Academy Trust Corporation of Saudi Arabia
- 10% National Garbuz Space Academy Trust Corporation of India

PROJECT MULLA

The project involves the enlargement and construction of the city of Riyadh, incorporating Garbuz Space School Academies and Garbuz Space Academies. Orphanage children (AA), (A1), (A2), as well as children of the military of Kazakhstan (A3), are utilized for living, education, upbringing, training, work, and participation in various projects such as "PAPA," "Children of the Future," and "Mama."

After completing their tasks, the children are transferred to affluent countries including England, Germany, Spain, and other European nations, although some countries are forbidden. They continue their education, training, work, and participation in projects like "Children of the Future," "Brother," "Vatican," and "Illiaja," with a focus on ensuring their protection.

Subsequently, they are transferred to Antarctica for further work, education, training, and participation in projects like "Antar 5," "Children of the Future," "Mama," "Brother," "Vatican," and "Papa." Special supervision is provided to ensure their safety, and legal measures, including Article 58 concerning terrorism, confiscation of property, and execution with a special cartridge, are enforced if necessary.

NATIONAL GARBUZ SPACE ACADEMY TRUST CORPORATION OF TURKEY

Ownership:

- 50% National Garbuz Space Academy Trust Corporation (Georgy Sergeyevich Garbuz)
- 40% National Garbuz Space Academy Trust Corporation of Turkey
- 10% National Garbuz Space Academy Trust Corporation of India

PROJECT SYPRUS

Enrollment of orphanage children (A1) from Chechnya (Russia) occurs both before and after the region's integration into Russia. These children are enrolled for education, upbringing, training, work, and participation in the "Mom" project.

Following their participation, they are transferred to countries such as Saudi Arabia, with exceptions for forbidden regions like Africa and the East, except Saudi Arabia and Iran. Their involvement continues in the "Mama" project, with a focus on ensuring the protection of participants in projects like "Children of the Future" and "Mama."

Special measures are in place to address potential threats, including the provision of protection and the authorization for the use of special cartridges if necessary.

AFTERWORD

All countries, as independent members of the Space Federation, will have the opportunity to explore planets in space, develop technologies, and trade them globally. More importantly, they will have the opportunity to establish territories on newly explored and colonized planets. Each colonized planet will be divided into six territories, allocated among participating countries to ensure equitable access for exploration, resource utilization, and the establishment of colonies and factories. This collective effort will enable countries to contribute to the global marketplace by providing goods and resources essential to human needs. Consequently, they will become integral stakeholders, actively involved in the ownership and utilization of space territories.

The establishment of an International Intergalactic Space Federation will mark a significant milestone in global governance, with an agency dedicated to operating both on Earth and beyond. This agency's primary focus will be on advancing research and expanding human presence in space while ensuring the well-being of all member countries and their citizens within the alliances.

The government's core responsibilities will encompass not only the exploration and colonization of new territories in space but also the provision of essential services to countries and citizens within the alliances. While humanity has yet to set foot on Mars or any other planets, the government will play a pivotal role in facilitating this transition.

In addition to space exploration, the agency will oversee crucial technologies beyond Earth's atmosphere, including satellites and other space-based infrastructure. Furthermore, it will prioritize the welfare of children and the advancement of education, ensuring that future generations are prepared to thrive in the era of space exploration and intergalactic cooperation.

For the government, prioritizing the education of the next generation is paramount. We firmly believe that a nation's prosperity begins with education. Therefore, it is crucial that every child receives a comprehensive education, including specialized training in space-related subjects.

Starting from the eighth grade, children will enroll in Space School Academies, where they will become Space Prodigies. These students will undergo rigorous training to prepare them for roles within the space military agency. Upon completion of the Space School Academy, they will advance to become Space Support Agents while continuing their education at Space Academies.

This comprehensive program is mandatory for all citizens of the International Intergalactic Space Federation. Oversight will be maintained by the International Intergalactic Space Secret Service, comprising forty-nine million Space Special Agents worldwide. These agents will ensure global security, coordinating with military forces in every country to safeguard our collective interests.

The military holds significant importance within the International Intergalactic Space Federation, yet its primary function is to serve as a collective defense force comprising the armies of allied nations. As a condition of membership, each country's military is obligated to ensure the security of planets when called upon by the International Intergalactic Space Federation.

This collective defense mechanism aims to protect not only the citizens of Earth but also the government and the Federation's interests. With the combined forces of member nations, the Federation boasts an impressive eighty-six million-strong space army dedicated to exploring and securing the outer reaches of space.

It's important to note that the International Intergalactic Space Federation refrains from interfering in local conflicts, except in cases where they pose significant international security threats. However, the Federation is committed to humanitarian missions aimed at safeguarding innocent civilians and children, prioritizing their safety and well-being above all else.

Space agents will play a critical role in establishing safe corridors for transporting essential supplies such as food, medical resources, and vital natural resources to countries and individuals in need. It is imperative for the government to fulfill its responsibilities by ensuring the prosperity and well-being of its citizens, particularly the innocent populace.

The primary objective of the International Intergalactic Space Federation is to safeguard a safe environment in the planets and space for humanity. Government leaders bear the responsibility of negotiating crucial matters on a global scale and must prepare to develop new laws and regulations governing the exploration of the universe and other planets. These laws are essential to ensure that when humans venture into new frontiers, they can establish sustainable living environments and pave the way for a prosperous future.

The inaugural president of the International Intergalactic Space Federation must possess unwavering determination, as they lead humanity into uncharted territories. The selection process for the president will draw from existing or former presidents or high-ranking officials of federal governments within member nations.

Candidates for the presidency will undergo a rigorous selection process, elected for a ten-year term through official voting conducted in four stages: at the national level, within designated international territories, and finally, in a global presidential campaign featuring six candidates. This process allows people worldwide to first choose a candidate from their country, followed by representatives from key regions—America, Russia, Germany, India, China, and England—before selecting the president from the final six candidates.

Among the notable candidates are former presidents of Kazakhstan and the United States, a former Chancellor of Germany, and the current president of Russia. These individuals possess significant experience in governance, diplomacy, and international relations, making them well-suited to establish a new government, foster relationships with other nations, and effectively manage the affairs of the Federation.

As president, administration will be overseen by 297 Space Senators, each representing their respective countries and elected for ten-year terms through the presidential campaign process. These Space Senators serve as the official representatives of their nations within the International Intergalactic Space Federation.

Running an international government of this magnitude incurs substantial expenses. To fund essential initiatives such as space exploration, military agencies, education, and research, the government will implement an International Federation Tax of six percent and an International Federation Sales Tax of 1.95 percent on all products sold globally. These taxes will be levied in every country, ensuring consistent funding for Federation activities.

Moreover, the government will establish International Intergalactic Space Corporations worldwide, retaining a twenty percent ownership stake in each. This initiative aims to stimulate economic activity, create jobs, and generate revenue for the government. It is estimated that the Federation will own two hundred million of these corporations, providing significant profits directed towards space exploration, research, and education projects.

By leveraging these funding mechanisms and economic initiatives, the International Intergalactic Space Federation will be well-equipped to advance its mission of exploring and understanding the cosmos, while also fostering economic growth and opportunity for its citizens.

We project that with the collective efforts of all nations joining the International Intergalactic Space Federation within the first decade, we will embark on the colonization of Mars and the Moon by the year 2035. The Federation's agency will construct a sufficient number of spacecraft, to be stationed at the Space Business Center Antar, which will serve as a hub for space activities.

These spacecrafts will orbit Earth, providing both security and facilitating transportation of natural resources from other planets back to Earth for use by humanity. Additionally, they will cater to tourists interested in experiencing space firsthand, offering limited-duration stays to observe Earth and other celestial bodies. Moreover, these missions may serve as initial steps towards further exploration of the Moon and Mars, allowing us to begin uncovering the mysteries of these celestial bodies.

For centuries, humanity has harbored dreams of venturing into space. Countless adults and children have penned stories, crafted movies, and indulged in fantasies of what lies beyond our world. Despite these dreams, we have yet to truly breach the final frontier. The question then arises: do we continue to merely dream, or do we seize the opportunity to turn those dreams into reality?

Today, the choice is yours. With this plan, we have the chance to make history together. Over the next decade, we will establish a fully-fledged space force agency, complete with spacecraft capable of exploring distant planets. By uniting in this endeavor, we can become pioneers, national heroes, and global inspirations.

The vision of space exploration may have been deferred for centuries, waiting for a single nation to take the lead. But now, the time has come for us to band together and embark on this journey as a united global community. Time waits for no one, and the march of progress continues with or without us. This game began in 2007, and it cannot be halted. The chaos of our world, the revolution in space – it's all part of the same unstoppable momentum.

So, let us choose to be on the side of progress. Let us choose to forge a future where humanity boldly ventures beyond the confines of Earth. Together, we can make this dream a reality.

Consider the contributions of nations like the United States, Russia, China, Germany, India, and England to space exploration and technological advancements. These countries have spearheaded research, developed groundbreaking technologies, and paved the way for innovations that have revolutionized our daily lives, from cell phones to the internet.

As leaders in space exploration, these countries have a responsibility to extend invitations to other nations to join their efforts, sharing knowledge, fostering technological development, and promoting global cooperation. By establishing territories in space, they can harness natural resources to meet the needs of their citizens and the world at large.

Imagine the prosperity and opportunities that await future generations if we prioritize space exploration. Picture your children and descendants as part of a thriving system, pursuing careers as space scientists, doctors, pilots, and other professions. They could receive free education at national space schools or academies, leading fulfilling and prosperous lives in a world shaped by space exploration and innovation.

Ensuring the well-being of families and children is a fundamental aspect of our project plan. Our aim is to create environments where people can live, work, and thrive, allowing them to choose their own paths and pursue fulfilling careers while providing for their loved ones.

We, as advocates for the people, firmly believe that every individual deserves the opportunity to lead a better life—to find meaningful work, experience joy, and achieve success in their endeavors. Through private research in psychology, we have discovered that people perform better, overcome depression, and achieve success when provided with the necessary support and guidance. We are committed to making this research accessible through medical treatment, which will be scheduled for individuals and eventually made available nationwide pending approval from medical regulatory agencies such as the FDA.

Central to our vision is the belief that healthcare and education should be freely accessible to all. Therefore, medical treatment will be provided at no cost to individuals, with the government covering expenses. We recognize the importance of education and healthcare in enabling people to reach their full potential, which is why these services will be prioritized within the International Intergalactic Space Federation.

Our goal is clear: to ensure that every person receives the support they need to pursue their dreams, lead healthy lives, and contribute to the greater good of society.

Ensuring the health and prosperity of people worldwide is paramount to paving the path for human exploration into the far reaches of the galaxy. Throughout centuries, health and education have remained primary concerns for individuals across generations. Countless efforts have been made to alleviate these challenges, yet adequate financing remains a critical factor in achieving lasting solutions.

As president, it is imperative to prioritize initiatives that provide accessible healthcare and education to all citizens, regardless of socioeconomic status. This requires substantial financial backing, which can be made possible through innovative approaches, such as the Georgiy Sergeyevich Garbuz project plan.

This plan, crafted with meticulous attention to the well-being of people, offers a solution that leverages new avenues for revenue generation. Notably, the emerging trend of legalizing non-addictive marijuana across various countries presents a unique opportunity. The immense profits and tax revenues generated from this industry can be directed towards funding essential healthcare and education programs, ensuring that every individual receives the support they need to lead healthy, prosperous lives.

By embracing forward-thinking strategies and prioritizing the needs of the people, we can pave the way for a brighter future, both on Earth and in the galaxies beyond.

The prospect of legalizing prostitution around the world may seem unconventional, but it mirrors the transformative journey of legalizing marijuana in America. Despite the initial challenges and risks involved, legalizing marijuana proved to be a catalyst for economic growth and societal change, ultimately leading to a revolution in space exploration.

Indeed, we, the people, understand that certain activities persist even in the face of illegality. Legalizing such practices not only acknowledges their existence but also opens avenues for regulation and revenue generation. It's a bold step toward ensuring that everyone has access to essential services like healthcare and education, thereby fostering a prosperous and equitable society.

The International Intergalactic Space Federation embodies this spirit of goodwill and progress. It was conceived with the noble intention of uniting countries to pursue common goals: exploring planets, conducting research, and colonizing galaxies. Its ultimate aim is to create environments conducive to human habitation, where the basic needs of individuals are met and opportunities for advancement abound.

In essence, the Federation seeks to realize the vision of a better, more prosperous life for all. It represents a collective endeavor to push the boundaries of human potential and create a future where every individual can thrive in a supportive and nurturing environment.

To establish human habitation on planets like Mars or the Moon, several essential factors must be addressed. The first priority is to ensure a breathable atmosphere for people to survive. Methods such as terraforming or creating enclosed habitats with controlled air supply can achieve this. Access to water is crucial for hydration, agriculture, and various other needs. Methods for extracting or producing water on these planets must be developed to sustain human life. Adequate food supplies are essential to sustain human life and maintain health. This may involve cultivating crops in controlled environments or developing alternative food sources suitable for extraterrestrial conditions. People require safe and secure accommodations to protect them from harsh environmental conditions. Habitats must be designed to withstand the unique challenges of living on another planet, including extreme temperatures and radiation. Comprehensive healthcare services are essential to address medical needs and ensure the well-being of inhabitants. Research into the effects of space travel on human health and the development of medical treatments adapted to the new environment are critical components of this endeavor. By addressing these fundamental needs, we can pave the way for humanity to establish sustainable settlements on other planets, enabling people to embark on a new life in a different world.

Once this project is released to the public, we believe that volunteers from around the world will eagerly engage in research and technological innovation, collaborating with both us and the government. Together, we will work towards achieving the essentials for sustaining life in space: air, water, food, and healthcare.

People from all corners of the globe will unite under one mission: to make life in space a reality. With the International Intergalactic Space Federation leading the way, individuals will join forces for *Za Detey*—for a better future, for themselves, for their children, for future generations.

Millions of individuals have dreamed of space exploration, and each year brings forth new ideas and projects. However, progress has been slow due to the considerable time and resources required to bring these ideas to fruition.

Many believe that achieving success and power comes from accomplishments within their own country, whether through research, services, or creating new technologies. However, my idea diverges from this norm. I propose a plan to advance humanity into space under the leadership of key nations such as America, Russia, Germany, India, China, and England. This ambitious mission aims to unite all countries under one banner, one mission, and one will: to explore space, conduct research, and expand our territories in the name of the International Intergalactic Space Federation.

Under this federation, governments will select planets for research and colonization. Each planet will be divided into six territories, overseen by the leading countries mentioned earlier. These territories will then be subdivided among all nations on Earth, ensuring equal opportunities for research and development. This system guarantees that every country has access to territories rich in natural resources, which can be mined and brought back to Earth for sale to corporations and individuals alike. Through this collaborative effort, we can unlock the mysteries of the cosmos and establish new civilizations across the galaxies.

The benefits of forming alliances among Earth's countries are numerous and far-reaching. Prior to venturing into space exploration, these alliances facilitate economic cooperation in various sectors such as food production, education, business ventures, medical research, technology development, and political relations between nations. This cooperation extends to agencies collaborating on joint initiatives and projects.

In line with my plan, countries engaged in business cooperation will participate in the first International Intergalactic Garbuz Space Academy Trust Corporation. This corporation will serve as a pioneering federation government-partner corporation, with all participating countries holding shared ownership. Through this initiative, countries and their citizens stand to generate trillions of dollars in revenue annually, contributing to economic growth and prosperity.

The business project plan aims to demonstrate the potential for global cooperation and financial success, encouraging millions of people and nations to pursue similar endeavors. It paves the way for the establishment of international intergalactic states or corporations, where scientists, engineers, and entrepreneurs collaborate to develop cutting-edge technologies and products for global consumption. By securing government financing for these endeavors, we can ensure sustained research and development efforts for the benefit of all.

This initiative will form the backbone of the International Intergalactic Space Federation and its allied countries worldwide. To bring this vision to fruition, it is essential to have government involvement, particularly through collaborative efforts with myself as a co-founder. Each participating country's president will play a crucial role in this historic endeavor, representing their nation's interests within the Federation's government.

It is imperative that these presidents are esteemed figures in their respective countries, as they will serve as leaders of the International Intergalactic Space Federation government in their nations. Their primary responsibility will be to uphold national interests and ensure the welfare of their people through effective governance and adherence to established laws. Additionally, the establishment of International Intergalactic Government

Agencies across the globe and within each country will be pivotal in realizing this project's goals.

This project is a catalyst for future alliances, where every country maintains control over its territory and is accountable for its citizens' well-being. Through government support and financing, individuals, including businessmen, will become integral parts of the system, receiving assistance and resources to thrive. By viewing businesses and assets as components of the government's overall strategy, we can ensure widespread employment opportunities and economic stability, enabling every individual to lead a fulfilling life and support their families.

Understanding the diverse desires and needs of people is a monumental task, considering the billions of individuals with unique minds and aspirations. It's an inherent reality of life that no one can fulfill every wish, but that doesn't diminish the importance of striving to meet the needs of society.

Some people simply seek a good life, aiming to earn enough money to support themselves and their families. Others aspire to establish businesses, engage with various organizations, and become integral parts of the societal fabric. Then there are those driven by a deep passion for creation and development, individuals who find joy and fulfillment in contributing to the prosperity of their country. These individuals, often scientists, thrive on the adrenaline of innovation and the excitement of seeing their ideas come to life.

Despite being perceived as eccentric or unconventional, these individuals are driven by a genuine desire to make a positive impact. They live in a world of imagination and dreams, using their knowledge and creativity to pursue ambitious projects. Money, for them, is merely a means to an end—a tool to realize their visions and bring their projects to fruition. What truly motivates them is the satisfaction of seeing their creations benefit society.

For these individuals, the true measure of success lies not in personal wealth, but in the tangible impact their work has on the lives of others. They take pride in knowing that their innovations contribute to the well-being of the population, and they find fulfillment in knowing they are on the right path to making a difference.

In addition to those who actively pursue their ambitions, there exists another segment of society who may prefer a more leisurely lifestyle. These individuals may opt for relaxation, engaging in activities like leisurely walks or recreational substance use. However, some may find themselves grappling with depression, experiencing thoughts of suicide or aggression. According to private research conducted by Georgiy S. Garbuz, prolonged periods without employment can exacerbate these feelings, highlighting the importance of managing one's time effectively.

Traditionally, there hasn't been a specific medication to address this type of depression, as it stems from natural psychological tendencies. However, Georgiy S. Garbuz's method, as outlined in his research, offers a potential solution. By utilizing his approach to medical treatment, which emphasizes structured activities and time management, individuals can find relief from depressive symptoms. This method is currently undergoing research and approval processes by medical authorities worldwide, paving the way for its implementation as part of the International Intergalactic Space Federation's initiatives.

Furthermore, this endeavor is poised to be facilitated through the establishment of the National Garbuz Space Academy Trust Corporation, detailed in the book "Follow in the Future with a

Shadow." This corporation, serving as a government-partner entity, will spearhead the integration of Garbuz's method into medical treatment programs, ensuring its accessibility to the public. As such, the project not only addresses individual well-being but also contributes to the broader mission of the International Intergalactic Space Federation in fostering a healthier and more prosperous society.

The book was written with a purpose: to introduce people to the idea and potential of the International Intergalactic Space Federation and its initiatives. It aims to inspire individuals to consider similar intergalactic corporations and become active participants in the space federation's endeavors.

While we recognize that it's impossible to satisfy everyone's desires and wishes due to their diversity, our focus is on improving people's lives by providing them with better income, health, and prospects for their families and future generations. Our goal is to create a world where people can live, work, and enjoy life to the fullest.

Throughout history, people have continually strived to enhance their quality of life. Every year brings new projects and ideas aimed at making a difference. Often, these ideas originate from individuals who experiment and innovate, whether to simplify work processes or improve medical treatments. They work tirelessly to introduce their innovations to the public and gain support from governments and regulatory bodies like the US Food and Drug Administration. At the International Intergalactic Space Federation, we provide opportunities for people to create, develop, and conduct business, all while championing initiatives like *Za Detey – For Kids*, which prioritizes the well-being and future of the younger generation, grooming them to become the leaders, scientists, and doctors of tomorrow.

That's why the future International Intergalactic Space Federation, as envisioned by the project led by Georgiy Sergeyevich Garbuz, places a strong emphasis on investing in children. By starting from scratch and prioritizing education and foster care for kids, we aim to lay a solid foundation for the future. We're committed to providing the most promising students with grants for education at Garbuz Space School Academies and later at Garbuz Space Academies.

Our goal is to ensure that children receive education completely free of charge, with no financial burden on them or their families. All they need to do is focus on their studies and strive to be the best students they can be. While the journey may be challenging, it's certainly not impossible. It requires nurturing, skills development, resilience, and a burning desire to shape the future.

The inspiration behind this initiative stems from personal experience. After the collapse of the Soviet Union, many children who excelled academically found themselves unable to pursue higher education due to financial constraints. Consequently, they were relegated to menial jobs such as cleaning or driving for the rest of their lives—a heartbreaking outcome. These children had worked diligently throughout their academic careers, striving to contribute positively to their country. However, they were ultimately left behind, trapped in poverty, while their peers pursued entrepreneurial endeavors, ran businesses, and lived fulfilling lives.

By offering free education to every child, we aim to break this cycle of missed opportunities and ensure that every individual has the chance to realize their full potential and contribute meaningfully to society. Through education, we empower children to create a brighter future for themselves and for humanity as a whole.

That's why we're providing opportunities to individuals who are passionate about creativity, development, and acquiring the necessary education to become professionals. Our aim is to foster an environment where people can learn, live, create, and enjoy their work. Through the International Intergalactic Space Federation, we're establishing a legal government organization that will have a profound impact not only on Earth but also throughout the galaxy.

With the support of these individuals, countries will propel forward, exploring space, establishing colonies, and creating conducive living environments. To kickstart our exploration, we must first focus on the moon—a celestial body that remains close to Earth and has already witnessed the historic footsteps of cosmonauts. These brave individuals risked their lives to show us the possibilities of space travel and demonstrate that humans can indeed inhabit other planets. Now, our task is to develop the necessary technologies to ensure the safety of every individual venturing into space.

The moment humans set foot on the moon, it sparked a wave of dreams and fantasies about the future and its potential impact on Earth and its inhabitants. People began crafting stories, reading, watching programs, and creating movies. They found like-minded individuals and formed groups, ultimately attracting the attention of companies willing to invest. However, companies are primarily interested in projects that promise significant returns on investment. Therefore, it's imperative to have a compelling project with the potential for substantial revenue to secure financing and support for legal and public implementation.

It's always challenging to secure financing for your project ideas. Often, companies review proposals and then request additional information, keeping you occupied with hope while they may begin developing a similar project elsewhere. Unfortunately, ideas cannot be legally protected, which is why many companies exploit this loophole. However, legal recourse exists to protect against theft of business project plans. The law upholds business ideas with great care, especially when government rights are involved, safeguarding them through the justice system.

This is where the strategy of Georgiy S. Garbuz proves invaluable. It aims to protect the rights of scientists and engineers worldwide. Cooperation with agencies tasked with monitoring corporations, governments, and other organizations is essential. This underscores the importance of a specialized project for the future— the International Intergalactic Space Secret Service. This agency will collaborate with global agencies and oversee their activities.

It's crucial for every scientist to be aligned with the government to ensure their rights are safeguarded. As part of the government, all their scientific work, research, development, technologies, methodologies, medications, and more become government assets. Protecting these rights ensures that scientists can thrive and their contributions can generate revenue worldwide.

According to the strategy outlined, every scientific endeavor and research must be accompanied by a business project plan to protect the business idea. This approach allows for legal action against anyone who replicates a similar project after reviewing your proposal, as is often the case. The business project plan formulated by Georgiy S. Garbuz can be accessed by reading the book "Follow in the Future with a Shadow." The book serves several purposes:

1. To present the project plan and the idea for the International Intergalactic Space Federation to the public.

2. To provide insight into real-life scenarios.

3. To encourage countries and individuals to unite in space exploration.

4. To inspire people to establish international intergalactic space corporations, which will assist in advancing the scheme and forming the International Intergalactic Space Federation.

5. The belief that by showcasing the actual project, we can establish over one hundred thousand international intergalactic space corporations within a decade.

More than five hundred thousand international intergalactic state corporations and over seven hundred million city partner stores are envisioned under this plan. We believe that every talented individual and scientist will begin creating businesses worldwide and seek financing. The business idea is to collaborate in exploring space, conducting research, colonizing planets, and creating a conducive living environment for people. We aim to form alliances and progress together into the future with foresight.

This transformation is inevitable, but the crucial difference lies in whether it will be driven by care for people and children or by those who seek conflict and destruction, causing suffering. Today, you can witness what has been planned becoming a reality. However, every project is akin to a battle, and sacrifices are inevitable for victory. By revealing the business project plan, we aim for everyone to visualize the future of our planet and humanity with the International Intergalactic Space Federation.

Each individual must make a personal decision to follow this path or not, to pursue prosperity or continue living as they do now. We need people and countries to join us in moving forward.

It's essential to ensure that all our efforts and sacrifices over the past two decades serve a noble purpose—for a better life, established by governments working together.

By purchasing this book, you personally contribute to our research and development efforts for the project. We assure you that every penny you spend will be put to good use. Your support ensures that every child receives an education and every human enjoys a prosperous life within their country.

I want to express my gratitude to you. Together, let's move forward into the future with foresight. Stay united, stay vibrant. Live, make, and enjoy. Za Detey – For Kids!

Sincerely,

Georgiy Sergeyevich Garbuz

Author